TREES OF INDIA

GW00384941

The WWF-OUP Nature Guides are simple and informative books
on our natural environment.

Other books in the series
Butterflies of India
Fishes of India
Seashore Life of India

WWF-India is one of the largest non-governmental organizations in the
country working for the conservation of bio-diversity and natural habitats.
With a large programme base in species, forests, freshwater habitats,
climate change, eco systems, livelihoods, and environment education and
policy, WWF-India engages with multiple stakeholders—local
communities, teachers and students, state and central governments,
industry and civil society organizations—to ensure a
living planet for future generations.
For more information, please visit www.wwfindia.org

TREES OF INDIA

PIPPA MUKHERJEE

Illustrated by
POONAM DESAI

OXFORD
UNIVERSITY PRESS

YMCA Library Building, Jai Singh Road, New Delhi 110 001

Oxford University Press is a department of the University of Oxford.
It furthers the University's objective of excellence in research, scholarship,
and education by publishing worldwide in

Oxford   New York
Auckland   Cape Town   Dar es Salaam   Hong Kong   Karachi
Kuala Lumpur   Madrid   Melbourne   Mexico City   Nairobi
New Delhi   Shanghai   Taipei   Toronto

With offices in
Argentina   Austria   Brazil   Chile   Czech Republic   France   Greece
Guatemala   Hungary   Italy   Japan   Poland   Portugal   Singapore
South Korea   Switzerland   Thailand   Turkey   Ukraine   Vietnam

Oxford is a registered trade mark of Oxford University Press
in the UK and in certain other countries.

Published in India by
Oxford University Press, New Delhi

Text © World Wide Fund for Nature-India 2008
Design and Concept © Oxford University Press 2008

The moral rights of the authors have been asserted
Database right Oxford University Press (maker)

First published 2008
Second impression 2009

ISBN-13: 978-0-19-568798-9
ISBN-10: 0-19-568798-1

Typeset in Aldine401 BT
by Pealidezine, Vasant Kunj, Delhi 110 017
Printed in India at Thomson Press (India) Ltd., New Delhi 110020
Published by Oxford University Press
YMCA Library Building, Jai Singh Road, New Delhi 110 001

CONTENTS

PREFACE

The purpose of this book is to describe and illustrate, simply, some of the beautiful, eye catching, and useful common trees that we see in India in our everyday lives. These can be seen by road sides, in gardens and parks, or all around the countryside.

I have tried to use simple language and have only explained a few words that might not be in common usage through additional notes in the text. A simple glossary has been included at the end of the book. This is meant to create interest and not to confuse either a child or a non-botanist with complex scientific terms.

I hope that the book will create an awareness and interest, in both young and old alike, and give people an idea of the never-ending, wonderful changes that occur in nature. I have also tried to make a list of the most important uses of trees for man, animals, birds, and, most vitally, for the environment.

I would like to thank Madhu Ramnath and Girija and Viru Viraraghavan for their invaluable help in botany and expert advice on trees in general; Mark Antrobus, for lending some excellent old books to me on trees and their uses; and Bob Granner, for his wonderful proof reading and advice on syntax, grammar, and English usage. I also need to acknowledge the following people for their help with the photographs of the new trees added to the book—Ian Lockwood, Suvir Mirchandani, Kuncelat Keystone Foundation and Bharath Sandur, Manish Chandi, Madhu Ramnath, and Jaap and friends from Auroville. Of course, the book would not have been complete without the immense care taken by Poonam Kurdekar (Desai), who repainted all the illustrations from the original book to make this volume complete. I cannot thank her enough.

I owe much of the updating of this book to Manjula who has worked for me for over twenty years and has learned from scratch to be an excellent gardener. She has also cooked for me and allowed me to continue my research without any domestic hindrances. Finally, I will be forever grateful to her for having given me not only many insights into the Tamil way of life, but also information about the use of trees in tribal and village culture.

I dedicate the book to my parents, John and Helen, who were responsible for my interest in nature. I would also like to recognize the work of the Palni Hills Conservation Council here. As a member of the Council's committee since 1985, I have been privy to much environmentally sound work by this NGO and I hope the Council will continue in strength over the next generation.

The following books and authors have been most helpful in the rewriting of this book: J.F. Dastur, *Useful Plants of India and Pakistan*; Sankar Sen Gupta, *Tree Symbol Worship in India*; E. Blatter and W. Millard, *Some Beautiful Indian Trees*; D.V. Cowen, *Flowering Trees and Shrubs in India*; C.S. Venkatesh, *Our Tree Neighbours*; C. McCann, *100 Beautiful Trees of India*; K.C. Sahni, *The Book of Indian Trees*; I. Pleiderer, *The Life of Indian Plants*; and R.N. Parker, *Forty Common Indian Trees*. For a very small amount of the up to date material, I have used various websites listed under the different trees. Most of the information, however, comes from first-hand experience during my thirty-five years in India.

Kodaikanal   PIPPA MUKHERJEE
March 2008

INTRODUCTION

When Mother Nature made the earth, she obviously had a very good plan for how each part of nature should fit together so it would work like a well-oiled machine. To show you what I mean, I will give you an example of a story in nature that is so strange, it sounds almost like fiction.

You may have heard of the Dodo. It was a large flightless bird that once lived on the islands of Mauritius and Reunion in the Indian Ocean near the coast of Africa. Because of man, it became extinct. Three hundred years ago, when ships anchored near these islands, the sailors would land and, while exploring, find the poor clumsy Dodos. As they were large and could not escape, the sailors killed them just for fun. After a very short time, the poor birds ceased to exist anywhere in the world. This was senseless cruelty since even the flesh of the bird was not particularly pleasant to eat.

But this bird had a purpose on earth which has been realized only recently. This was to help the seeds of a tree called the *Calvaria major* to germinate (grow) first into plants and then into full-grown trees. The seeds were eaten by the Dodo. While passing through its digestive system, which was extremely long, the hard outer covering of the seed was dissolved. So when at last it was passed out of the bird's body, the seed was ready to germinate. Since the disappearance of the Dodo so many years ago, no more Calvaria trees have grown. The trees that remain are very old and are now dying.

Scientists have been working hard to find another bird to do the job that the Dodo once did. It is now thought that the turkey is a possible substitute as it has the same kind of digestive system. Let us hope that although the Dodo cannot be restored to the earth, at least the Calvaria tree will not be lost.

Recent research has revealed that sailors alone were probably not responsible for the demise of the Dodos and it is more likely that domesticated animals such as pigs were responsible for destroying the Dodos' ground-level nests and eggs in their search for buried roots and other food.

Another more recent example of man's abortive attempt to control nature has become been even more problematic. In 1935, the cane toad, a large amphibian, weighing up to two or three kilos, was brought from the island of Hawaii to Australia. This was a vain attempt to combat Greyback beetles that were destroying the tropical sugar cane fields there. This toad that eats the larvae of the

beetle could not rid the cane fields of the beatles, as the larval period of the beetles did not coincide with the toads' breeding period when it needed the additional nourishment. So, the Australians who had the beetle to contend with, now also have the toad. Now there are millions of toads in northern Australia and each female can lay up to 35,000 eggs at a time. The toads are extremely poisonous and have poison sacs behind their heads, and a crocodile who has eaten a toad will die in fifteen minutes. Also Quolls, little animals closely related to the Kangaroo family, who are ferocious carnivores are severely threatened as they often eat toads. It is common in northern Australia for people driving home at night to try to eliminate toads crossing the road by swerving from one side to the other to crush as many toads as possible.

Through these examples it becomes clear that everything in nature is connected and thus there are many trees whose seeds need the help of the digestive systems of birds or animals to grow. You will notice that I have pointed out several such examples in this book.

The stories I have just told you show very clearly how everything on earth is related directly or indirectly to every other thing and how man can interfere with this relationship. Destroy one thing and everything else will be affected in one way or another. This balance is called the ecological balance and it is very important to maintain it for the well-being of the entire world.

If, for instance, all the trees in one forest were cut down for some reason and another kind of tree not usually grown in that area was planted, it would completely change the balance of nature. Insects, birds, and animals would either die or move elsewhere to find their normal food, and other insects, birds, and animals might or might not replace them. Similarly, the small plants normally growing beneath the trees would change for the same reason. People living in the forest would find that the trees from which they got their fruit were no longer there and that animals they hunted for meat had moved away.

Now we will talk about trees in general. The first thing that I want to say is that trees of the same kind have different habits and these depend on where they grow. For example, a tree growing in Mumbai may flower in January or February, whilst the same tree in Delhi or Uttar Pradesh may flower in August or September or some other month of the year. This is very important to remember when recognizing trees.

The soil the tree grows in and the air around it make a difference to the way that it grows. For example, when the air is polluted with exhaust fumes or carbon dioxide and other unpleasant gases, rather than being clean and fresh, and full of oxygen, the growth of the tree is severely affected. If a tree has poor soil and little fresh air, it will grow up stunted in the same way that you would if you lived in poor conditions and had little or no food. If people tell you that a tree loses its leaves in September–October and flowers in February–March, remember that it might not do exactly that.

This is the interesting part of nature, and this is what I would like every one of you to start watching out for. Towards the end of the book you will find a page which tells you some of the things to look for when recognizing or simply looking at a tree. This can be really exciting as there are so many things to see. I hope that it will encourage you all to be more observant duing your daily walks.

GENERAL DESCRIPTION OF TREES

All trees have trunks with which to support the spreading branches above. This trunk has a protective covering or bark which is essential for the well-being of the tree. If, for any reason, this bark is destroyed by animals or by man, the tree will die. This is because the bark protects the sap or blood of the tree which rises up under the protective layer or bark to supply the branches, leaves, and other parts of the tree with nutrition.

EVERGREEN OR DECIDUOUS

You may have heard the terms 'evergreen' and 'deciduous' used in connection with trees. In case you have not, I will explain them simply.

If a tree is evergreen, it means that all its leaves never fall at the same time, thus leaving the tree bare. They will fall, a few at a time, and the new leaves will grow in the same way. You may notice the new leaves as they are often a brighter green than the older ones. All this happens very gradually. Also, the leaves of an evergreen tree are usually shinier and thicker than those of a deciduous tree.

If a tree is deciduous, it means that the tree will lose all its leaves once, or in some cases, twice a year and become completely leafless. Later, all the new leaves will grow and open at the same time.

Finally, because of the climate they live in, there are trees that are neither deciduous nor evergreen. The Badam tree is one of them. These trees are called semi-deciduous, or nearly evergreen. You will notice this later in the book when we get on to the section on individual trees.

ROOTS: These differ from one tree to another in size and shape, but basically, they do two jobs. They anchor the tree and stop it from falling down, and they feed the tree by sucking food from the earth and allow nutrients to be pulled up to feed all the branches and other parts of the tree.

The water and mineral salts needed by the tree are often taken from the rich shallow soil near the surface. It is therefore the job of the roots at the top to feed the tree while the deeper roots act as anchors. Sometimes, however, when water is difficult to obtain, the roots have to grow deep down to reach the necessary moisture. Some trees are so large that they need to support the base with spreading roots high above the ground, like pillars in a cathedral or temple. These are called butteresses.

LEAVES: The colour and shape of leaves, and the way they grow differ from one type of tree to another. By studying these one can quickly recognize each one. In hot countries, it is quite usual to see young new leaves that are reddish brown or pale pink in colour. This is a protection against strong sunlight which could otherwise damage them. But, most full-grown leaves are green as they contain chlorophyll. This is the substance which, in the presence of sunlight, helps the plant to change carbon dioxide and water into sugars. By doing this, it feeds the plant or tree and gives us oxygen as a byproduct.

AGE OF TREES: This again depends very much on the conditions in which the tree lives. If, for example, a Pongam tree grew in a city with very poor soil, it would be likely to have a shorter life than the same tree growing in a village with plenty of space and good soil. Of course, there are some trees that are well-known for their long lives. The Banyan, the Peepal, and the Tamarind have been known for centuries and some of them have lived for a thousand years or more.

SPEED OF GROWTH: Most trees do not grow very fast. It is difficult to give growth measurements in feet or inches of how much a tree grows in one year. Again, their living conditions play an important role in their growth and also determine whether they have soft or hard wood in the trunk. Generally, the rule is the harder the wood, the slower the growth.

In some trees, there are definite scars left on the trunk at places from where old leaves have fallen. The rings on the trunk of the coconut palm, for example, clearly show how long the palm has taken to grow. But other trees do not have this feature and so it is very difficult to judge how fast they grow unless they are measured regularly. The important thing to remember is that many trees are very adaptable. This means that they can grow in places they are not native to. You will notice, when you read this book, that many trees originally came from other countries, but now grow well in India. Still, they often grow very differently from the way that they grew in their original homes.

Now I want you to think about the most important uses of trees. This will help you realize the vital role they play in our lives. Although you will notice that I often mention the uses of the wood of the trees that I describe in this book, felling of native trees is environmentally unwise. Unless it is absolutely necessary, it should not be undertaken. Natural tree cover in India has been reduced to a very small area because of illegal felling, which is very sad as we need trees for so many reasons.

What is certain now is that humans are responsible for the enormous amount of environmental damage that must now be redressed if the earth, as we know it, can continue its existence. Appreciating, planting, and caring for trees, may, in some measure, slow down the pace of destruction and it is the next generation who will have to clear up the mess that we leave behind.

Note: You will see that next to the botanical names of the trees in this book are short forms of names or initials such as L., Lamm., Aubl., Benth., and Forst. These were the eminent botanists who first described these trees for modern science. This information is included for the benefit of anyone who would like to know more about the scientific discovery of trees.

USES OF TREES

1. First, trees work very hard to keep the air we breathe clean and healthy. They are like sponges. Their leaves breathe in much of the poisonous unwanted carbon dioxide in the air and replace it with the oxygen that we need for breathing and healthy living. This system of absorbing gases on which all plants rely for their food is called photosynthesis. In this process, the plants, with the help of sunlight, water, minerals, and the green material called chlorophyll within the leaves change the unwanted carbon dioxide into food for themselves. When doing this, they release oxygen into the air, which is vital for our lives. At night, when there is no sunlight, the plant no longer makes food, so it does not release the same amount of oxygen. One is often told not to sleep with plants in one's room at night as they will use up all the oxygen. But, although a little photosynthesis may take place at night, when the plants rest, so little oxygen is absorbed from the air that little harm can come to the sleeper. You now see that what we do not need, plants require, and what they do not need, is essential for us humans and animals.

2. Tree roots dig deep into the earth and hold it together so that the rain and wind cannot wash or blow the soil away. This is very important as the earth only has a very thin layer—seldom more than one foot—of fertile soil covering it. If this were to be washed, blown, or worn away, leaving rock or sand exposed, then no plant would be able to grow, and the earth would eventually become a desert. The removal of this top-soil is called soil erosion. Scientists all over the world are trying to find ways to prevent it. One of the most important ways is by planting more trees, especially around fields, on river banks, and along coastlines.

3. Trees and plants also help to prevent floods. The roots keep the soil of river banks firm and do not let them crumble. Water is thus prevented from flooding fields and destroying farmers' crops, or entering villages and drowning houses, people, and animals.

4. Wherever there are groups of trees or forests, they attract rain. Through their leaves, trees release water vapour as evaporation into the atmosphere. When this vapour meets the cooler air high in the atmosphere, it turns into drops of water which then fall as rain. (The cooler the air, the less water vapour it can hold.)

5. Trees give us shade and protection from harsh weather. More importantly, crops such as coffee, cocoa, and tea are protected from strong winds, rain, or too much sunshine by trees that grow around and above them.

6. Trees give us beauty, colour, and greenery. This is something which we often forget and fail to appreciate.
7. Trees provide homes for many birds, animals, and insects. Each of these is important in keeping the balance of nature.
8. Trees give us food. (Think of how many fruits you eat!) Ropes, utensils, thatch, medicines, wood, paper, and so many other things we use in our homes everyday, or which are necessary for our health, are made from trees. These are only a few of the ways in which trees help us in our daily lives. Our very future depends on our ability to protect trees and plants. Now let's read a little about some of the common trees that grow around us.

Note: There are some words included in the text that you may not know. So, at the end of the book are simple descriptions or a glossary which will help you understand the meaning of the word. In some places, I have also put a 'footnote' at the bottom of the page describing a word or phrase.

AFRICAN SAUSAGE TREE

Bignoniaceae
Kigelia pinnata DC.

WHERE IT GROWS: This strange tree is native to tropical West Africa. It arrived in India when one fruit was washed up on the Mumbai shore in the 1800s. It became a popular ornamental tree because of its oddity and has now been planted along roads, especially in North India.

GENERAL DESCRIPTION: This is a medium to large semi-evergreen tree with a thick trunk covered in rough greyish-brown bark that flakes off easily. It grows best in the dry areas of India and can grow in more humid areas, but it does not produce fruit. The branches spread out in a twisted fashion to form a large canopy.

In English, this tree is also called Fetish tree and Cucumber tree; in Hindi and in Bengali, it is known as *jhar fanoos*.

LEAVES: The tree loses its leaves twice a year but is never completely bare. The leaves are large and leathery, and dark green in colour. They crowd at the end of the branches and lie opposite each other. The upper surface of each leaf is smooth, but there is soft down on the underside.

FLOWERS: The blossoms are large, showy, and may be crimson, maroon, or dark purple in colour. They hang like upturned cups on long rope-like stalks. Only a few open at a time and they bloom at night. The unpleasant smell that emanates from them attracts bats which pollinate the tree.* (The word 'pollinate', in case you do not know, means that birds, insects, and animals come to the flower to take nectar, and in doing so, rub some of the pollen off the stamen or male part of the flower onto themselves. When they fly onto another flower, they cause pollen to fall onto the female part of the flower, or pistil, so that the fruit can be formed.) By the next day, the blossoms fall, leaving a trail of old flowers underneath the branches. The flowering season is from May to July.

FRUIT: These resemble long, brown sausages or gourds, and are brownish-grey in colour and hang on long stalk-like cords, which can be up to two meters in length. Each curious fruit can weigh up to seven kilos and contains seeds inside the stringy pulp. Normally, these hanging cords are tough and do not break. The fruits remain on the tree until well after December or January.

USES: Since the leaves are not eaten by cattle, this is an excellent tree to plant in places that cannot be protected. The fruit can be cut and lightly roasted as a cure for rheumatism and as a dressing for ulcers. The bark is useful for treating dysentery. Some chemicals found in the tree are currently added to commercial products like shampoos and facial creams. The cooked seeds are sometimes eaten in times of food scarcity and the wood is hard, and often used for cabinet making. In some parts of Africa, the tree is considered sacred and people rear it with care.

*Flowers that are pollinated by bats usually have a strong, fruity fragrance at night. The bats are often described as smelling musky or 'batty' and love products that are fermented; bats that feed off nectar or pollen have a well-developed sense of smell, e.g., flying foxes.

AUSTRALIAN SILVER OAK

Proteaceae
Grevillea robusta A. Cunn.

WHERE IT GROWS: This tree is a native of Australia but is now grown widely in many parts of India. It is used as an ornamental tree as well as for shade in tea and coffee plantations. The trees protect growing plants from harsh sunlight, rain, and wind, and the crops, thus protected, are often of higher quality and produce better quality fruit.

GENERAL DESCRIPTION: A tall, straight, semi-evergreen tree, bearing no relationship to the true oak, except for the excellent timber it produces. It has a cone-shaped crown (the crown is an

Its other English names are Silk-oak, Silky Oak, or Golden Pine; local names are scarce. In Tamil Nadu, the tree is called *groulia* or *sivukkumaram*; and in Bengali, it is called *rupasi*.

umbrella-like canopy at the top of the tree) and short, symmetrical branches that are often covered in dark grey silky hair. It is fast-growing and prefers acidic soil. It sometimes becomes leafless for a very short period of time.

LEAVES: These are fern-like in structure, with an attractive feathery appearance. The fronds are dark green above and shining silver beneath. When the wind blows, the movement of the fronds draws one's attention to the tree as they flash silver in the light.

FLOWERS: The blossoms appear in great clusters on long spikes and are golden yellow in colour and tube-like in shape. Each tiny bloom is curved and the overall effect is to make the tree appear like a mass of brilliant colour against the bi-coloured leaves. The flowering season in most parts of the country is from March to April and in the higher altitudes, as late as October.

FRUIT: The pods are oblong and leathery. They open only after fire burns the pods to a certain temperature. The opened pod releases two flat, light, and winged seeds, which are wind-borne. The seeds germinate with ease.

USES: Apart from its use as an ornamental tree, its frond-like leaves are often used in flower arrangements. It is valued as a shade tree and the leaves make excellent green manure. The wood is tough and long lasting with a lovely grain. It is commonly used for carving panelling, making furniture, tennis racquets, and other fine wood work. Due to the cellulose in the wood, the trunk is often pulped to make excellent quality paper.★ The flowers produce delicious honey and from the bark, tannin (for tanning leather) and gum are extracted.

An additional use for the tree is as a boundary marker around properties or as shelter belts around fields. Shelter belts are commonly used by farmers to protect their soil from strong winds that may blow away the soil and create erosion. ('Erosion' is the process when soil is washed or blown away so that the field can no longer sustain a crop or vegetation.)

★Cellulose is a part or tissue within the cell walls of a plant.

BABOOL

Mimosaceae
Acacia arabica (L.)Willd

WHERE IT GROWS: This tree originated in India and Africa. It is very hardy and needs little moisture, and grows well in dry areas and also on wet land near the sea. It dislikes frosty weather and high altitudes, and is mainly a tree of the warm plains.

GENERAL DESCRIPTION: It varies in size from a small, spreading plant to a very large tree. It is evergreen and grows fairly slowly. In less suitable areas, it grows in an uneven, untidy way.

Its dark brown bark is cracked and uneven. The lower part of the tree is often covered with silvery white and dangerous thorns that prevent cattle from eating its lower branches. Often, the upper branches have no thorns at all. This tree is the favourite home of the striped palm squirrel, the weaver bird, and many other birds and animals, probably because the thorns protect them from predators.

Also known as the Indian Gum Arabic, it is called *kikar, babul, bhabhul,* or *babla* in Hindi, Marathi, and Bengali; *karuvalam or karuveli* in Tamil and Malayalam; and *nellatuma* in Telugu.

The butcher bird (Lahtora Grey Shrike), uses the thorns to store its food. It pierces and hangs the insects it catches onto them to eat later.

LEAVES: The leaves close when it is very hot to protect themselves from the sun and to preserve moisture. They are bright green and feathery.

FLOWERS: In some places, the flowering season is from June to October, but in others, the trees flower through the year. The flowers grow in groups of two or six and are faintly perfumed. Each flower develops on a tiny stalk growing out from between the leaf stalk and branch, and is golden yellow in colour.

FRUIT: These are leathery seed-cases (pods) containing about seven to twelve seeds, with each seed in a separate section. The Babool produces seeds that germinate after passing through the digestive system of animals or birds. This is because the seed has a very hard outer casing that needs to be dissolved by their stomach acids. Some gardeners boil or soak the seeds to get the same result.

USES: This tree is useful and important to village people. Its branches and leaves are cut, and used as food for cattle and goats. The heavy wood is excellent for making farm tools and is also used as fuel and for making charcoal. The gum, which comes from the trunk and oozes out of damaged parts of the bark from February to March, is used for making glue, sweets, and chewing gum. The thorns are used as pins and are still in common use in small rural Government offices.

The twigs of Babool are used for brushing teeth. Lac is obtained from this tree and from several other trees mentioned later. Lac insects are tiny 'parasites' (that is, organisms who get their food from the trees or animals on which they live). They live together in colonies on certain trees and produce a dark red liquid, which later hardens, and is then used to make varnish and many other things. Jewellers use lac to support the metalwork into which they set precious stones. When the work is complete, the lac is chipped away. Lac is also used for sealing packages and securing envelopes as sealing wax. When purified, lac becomes lacquer, which is used to decorate bowls and other utensils.

Some people believe that the spirit of a Muslim saint lives in the Babool tree and that if cloth is offered to it, it protects children and ensures that long journeys will be undertaken safely.

BADAM

Combretaceae
Terminalia catappa L.

WHERE IT GROWS: This tree is originally from the islands off the coast of Malaysia and the sandy coast of Malaysia, but it now grows widely in India. It likes sandy soil and tropical climates and grows well close to the sea.

GENERAL DESCRIPTION: It is a fairly fast-growing, tall, semi-deciduous tree, with a smooth grey bark. This tree is easy to recognize as its branches grow in layers, almost at right angles to the trunk, and begin fairly high up on the trunk. This tree requires quite a lot of space because its branches spread a long way outwards and as it gets older, it develops a broad spreading crown and sometimes, a buttressed base to support it. (A buttress is a broadened base at the bottom of the tree.) Since *badam* means almond in many Indian languages, you should not confuse this tree with the True Almond, which is a member of the rose family.

Also known as the Indian Almond, Malabar Almond, or Bengal Almond. Its Indian names are *deshi* or *jangli badam* in Hindi; *badam* in Marathi; *bangla badam* in Bengali; *badamuchetti* in Telugu; and *natvadom* in Tamil.

LEAVES: The large leaves are rough and leathery. Once or twice a year, they turn bright red and fall. The new leaves are a bright green but darken quickly as they age. When the leaves are young, they are often eaten by beetles and grasshoppers, so the young trees have to be looked after carefully.

FLOWERS: Small, greenish-white flowers grow in spikes at the end of the smaller stems or branchlets. They usually appear in March and April, and again in July and August.

FRUIT: The oval nut cases contain a nut somewhat similar in taste to an almond and can be eaten cooked or raw. They attract small children who love to throw stones at the fruits, much to the irritation of compound owners.

USES: In India and Malaysia, this tree is often planted around monasteries and in gardens for its beauty. If the nut is crushed, it produces oil that is very similar to almond oil. The nuts and oil cakes of this tree are also considered to be good food for pigs. Wood from this tree is used in construction. Ink can be made from the leaves and bark.

Tasar silk worms that weave their cocoons of silk thread from which tasar silk is then woven★ love to eat the leaves of the Badam. The trees are therefore often planted near silk farms to feed the caterpillars.

★Tasar silk is of a pale coppery colour and is more coarse than other silk fibres. But it has its own appeal for those who love the undyed, attractive weave. It is often used for furnishings, but also makes lovely saris and shawls.

BANYAN TREE

Moraceae
Ficus bengalensis L.

WHERE IT GROWS: This is probably the most common tree of India and it originated here. It was often planted along roadsides, but with the ever increasing traffic and widening of highways, it has, sadly, become unpopular because its aerial roots have to be cut. As a result, the tree loses its support and either falls or gradually dies.

GENERAL DESCRIPTION: Because of its size, its splendour, and the shade it gives, it is often planted (like the Mango tree) at the central meeting place in villages. There are over 600 different kinds of such fig trees that grow in tropical countries. From its branches hang roots which, when they touch the ground, take root and support the tree. Thus, the tree grows sideways, supported by hanging roots. In this way, some trees can cover a lot of ground. It is a tall, fast growing, evergreen tree. It

It is called *wad* or *vor* in Marathi; *ala* or *aalam* in Tamil; *bor* or *bargaa* in Hindi; *marri* in Telugu; and *bar* or *bot* in Bengali.

has shallow roots and a smooth, dark grey bark that peels in patches. It can grow to an enormous width and live for hundreds of years. The two largest Banyan trees in India are in Kolkata and Chennai. The name Banyan may have been given to this tree because for centuries, *baniya*s or merchants have taken shelter and traded under its canopy.

LEAVES: The leaves appear in February or March and sometimes, again in September and October. If a leaf is broken off, a white sticky liquid (sap) oozes out in all fig trees.

FLOWERS: The figs which contain the flowers grow in pairs just below the leaves and they look like cherries. Each fig contains a number of flowers, both male and female, so that the Banyan and all the other members of the fig family continue to produce seeds and increase in numbers. A wasp called fig wasp bores a hole and stays in the figs. If you cut open nearly ripe figs, you will see tiny fig wasps in it. You may find as many as eighty wasps in one fig. They are quite harmless and do not sting. They use the fig as a nest to lay their eggs and hatch their young. Some of the young wasps leave the figs carrying a little pollen from male flowers on their backs. This gets brushed onto female flowers on other fig trees and thus new seeds are formed.

FRUIT: The figs ripen between February and May, and in some areas, from August to September. They are the favourite food of many birds, flying foxes and other fruit-eating bats, and squirrels and certain kinds of insects. Often, seeds from the figs are dropped in strange places, for example, in palm trees. These seeds may grow and use the tree for support. They then grow so fast around the tree that they may kill the tree that has supported them. This is why they are known as strangling figs. Trees that do this are called epiphytes.

USES: The wood of the Banyan is not good, but the aerial roots are strong and are used to make tent poles. Rope is made from the bark and young roots. The leaves are used as plates and many parts of the tree are used in traditional medicine. The tree is regarded as sacred in many parts of India and in folklore, it is forbidden to cut a Banyan.

BELLERIC MYROBALAN

Combretaceae
Terminalia bellerica (Gaertn.) Roxb.

WHERE IT GROWS: This tree is native to India and is found in most of the deciduous forests of the country, except in the driest regions. It does not grow on high altitudes and any specimen found above 3000 feet is exceptional. It also grows in Myanmar (Burma) and Sri Lanka. This tree belongs to a large family of important trees and because of its handsome appearance, it is often planted along roadsides.

The other English name is Bastard Myrobalan; in Hindi and Bengali, it is called *bahera*; in Gujarati, *baheda*; in Marathi, *bherda* or *behda*; in Tamil, *tani*; in Malayalam, *thaani*; and it is known as *santi tandi* in Telugu.

GENERAL DESCRIPTION: It is a large, deciduous tree with a straight trunk, buttressed or enlarged at the base. Many horizontal branches grow out from the trunk, making a crown. The bark is dark grey, rough, and fissured with many fine, lengthwise cracks. Due to its size and girth, it is a very impressive tree when fully grown.

Sadly, now that roads in India are being widened to make way for increased traffic, many of these wonderful trees are being felled, leaving the roads exposed to the sun.

LEAVES: The leaves fall in December in dry areas and the tree remains bare until April. In humid or damp places, the leaves are lost for only about a month. The new leaves appear with the flowers and are an attractive coppery colour. They are large, leathery, and packed close together at the end of the branches. Oblong and wedge-shaped at the base, they have a broad central rib attached to a long leaf stalk or petiole. They are a glossy green above and greyish-green below. Just before they fall, they turn a brilliant red.

FLOWERS: In April and May, the flowers appear, springing out from among the leaves in long, creamy tassels. Each blossom is tiny and greenish-yellow in colour, and has a strong scent. This fragrance is honey-like, but I find it overpowering and quite unpleasant. The buds are globe-like and the blossoms look like tiny cups divided into five petals.

FRUIT: The fruit is a velvety grey ball, about one inch across, and contains a single seed inside a hard shell. The pulp around the shell is rather dry. The fruits ripen in December or January.

USES: The fruit is used in the herbal preparation, *triphala*. This is a combination of three fruits, one of which is another member of the same family and is called *Terminalia chebula*. The third fruit is the Indian gooseberry or *amla* (*Phyllanthus emblica*). Triphala is a tonic and a cure for many intestinal problems. '*Bahera*' is another drug prepared from the dried fruits of Belleric Myrobalan and is used as a cure for leprosy (Hansen's disease), fevers, and fluid retention.

The oil of the seed is used in soap-making. The seed kernels are edible and taste like almonds. The wood is yellow and subject to insect attacks, so can only be used after being soaked in water to make it more durable. In parts of South India, the tree is never cut as people think that the tree is inhabited by demons. It is believed that anyone who builds a house with the wood will be cursed.

BER

Rhamnaceae
Zizyphus mauritiana Lam.

WHERE IT GROWS: This tree originated in India and its neighbouring countries, and is very similar to its close relative from China. In fact, trees closely related to the Ber grow in Africa, Australia, and many other countries. Ber-growing states in India include Andhra Pradesh and Tamil Nadu.

GENERAL DESCRIPTION: This is a tree that generally grows wild and it grows best in dry areas. As it produces many seeds which grow easily, it is not grown in gardens because the seedlings have to be dug out to prevent too many growing and crowding out other trees.

It is a small to medium-sized, thorny tree, fast growing and deciduous. It often divides low down the trunk and always has many branches. On wastelands, it is often small and straggly, but in good conditions, it can grow into a tall, graceful tree with a spreading crown and branches that

Indian names for this tree are *bor* or *ber* in Hindi, Marathi, and Gujarati; *elandai* in Tamil; *keegu* in Telugu; and *boroi* in Bengali; the English name is Jujube tree.

sweep downwards. The bark is dark brown or black, very rough, and cracked. The young stems are pale green and covered with soft, brown hair.

LEAVES: The under-sides of the leaves appear to be almost white or pale brown because of the thick layer of soft hair covering them. In addition, they grow alternately in two rows up the zig-zag branchlets. The three main veins on the leaf are very clearly marked. When the leaf is in the bud stage, it is folded along the veins into four sections. At the base of the leaf stalks grow two short, sharp spines. These spines are nature's defence against grazing animals. In cultivated trees, the spines are either less frequent or totally absent.

FLOWERS: The main flowering season is September to December, but flowers often appear at other times. The flowers are five-petalled and greenish-yellow in colour. They grow in small groups on tiny stems and have an unpleasant smell.

FRUIT: The seed pods are fleshy and differ in shape and size, some being small and round, while others are oval and much larger. They contain one or two seeds in a wrinkled, pointed, hard case. The cultivated Jujube trees have much larger berries than those that grow wild. When ripe, they are a darkish red in colour and have a thin skin, dryish pulp, and a hard case containing two seeds.

USES: The berries are eaten as fruit and are a delicacy in early spring when they are sold in newspaper twists—a favourite with children of all ages. The fruits are also eaten by birds and animals, and the seeds germinate better after passing through the intestine of a bird or animal. Ber fruits are also used as souring agents or are preserved in syrup for future use.

The bark of this tree can be used for making tannin and the tasar silk worm feeds on this tree. It is also a favourite tree of the lac insect. The thorny branches of Ber are used for making fences and the wood is often made into agricultural tools or is used as fuel. The tree is revered by Sikhs and there is a Ber tree said to be over 450 years old near the Golden Temple in Amritsar.

BHENDI

Malvaceae
Thespesia populnea L.

WHERE IT GROWS: This tree belongs to India, Myanmar, and parts of Africa, tropical Asia, and the Pacific. It cannot live away from the sea or in places where it is cold. Its favourite home is in warm, tropical, coastal regions, preferably away from the hills, e.g., the Maharashtra coast.

GENERAL DESCRIPTION: It is a small to medium-sized tree and nearly evergreen. It grows fast but has a short life. It has a rough grey or brown knobbly bark and a crown of leaves which make the tree look like an umbrella.

Also called Umbrella, Portia, or Indian Tulip tree. It is called *parsipu* in Hindi; *dumbla* in Bengali; *bhendi-kejhar* or *bhendi* in Marathi; *paarsapeepala* in Gujarati; *galgaiavi* in Telugu; and *poovarasam* in Tamil.

When Captain Cook discovered this tree in Tahiti, he found that it was always planted near temples. Its Latin name 'Thespesia' means divine or holy.

LEAVES: The leaves generally fall in early spring but never all together. Since the withering leaves are bright yellow, they often make the tree look like it is covered in flowers. In some areas, it is virtually evergreen, losing only very few leaves at any one time. The leaves are heart-shaped and pointed with conscpicuous veins.

FLOWERS: These usually appear in the winter but many trees have a few flowers throughout the year. The blossoms are five-petalled and lemon-yellow in colour, and are set in a green cup. The petals are crinkled and tissue-paper thin. They look as if they need to be ironed! Inside the flowers are patches of deep maroon and purple, and as the blossoms fade, they turn from orange-pink to purple.

The name Bhendi is given to this tree because its flowers are very similar to those of the vegetable plant, *bhindi* or okra, commonly known as Lady's Fingers.

FRUIT: Small, bell-shaped seed-cases remain on the tree for a long time, first turning brown and then black, and contain oval, silky-white seeds.

USES: It is commonly planted on the sides of roads in cities to provide shade and beauty. The wood is water-resistant and durable, and is often used for building boats, furniture, carts, and in turnery, which is the art of making bowls out of wood using a wheel similar to that used by potters. From the bark, string is produced for making sack cloth and rope, but this is not so common in India. Medicines are produced from many parts of the tree. The fruits, leaves, and roots are used to cure skin diseases and the bark is used as treatment for diarrhoea. The seeds produce oil that villagers use for lighting lamps and the leaves are excellent fodder. Tribals often eat the buds and flowers, either cooked or raw. If the pods are crushed and applied to the forehead, they provide relief from migranes.

CANNON-BALL TREE

Lecythidaceae
Couroupita guianensis Aubl.

WHERE IT GROWS: This tree originated in French Guiana and South America, but can live happily in all tropical countries, and prefers moist, low-lying areas. It is a close relative of the Brazil Nut tree.

GENERAL DESCRIPTION: It is a large, fairly fast-growing, deciduous tree with a thick, straight trunk and a rough, grey-brown bark. Branches start growing fairly high up on the trunk. The tree often looks untidy as the branches grow unevenly.

In Marathi, this tree is called *kailashpati*; in Hindi, *shivalingam*; and in Bengali and Tamil, *naglingam*; it is also called Monkey Pots in English.

LEAVES: This tree sheds its leaves several times a year for a short period. It is common to be showered with yellow-green leaves if one walks under it. When the leaves fall, it only takes a few days for new ones to grow, so the tree is never completely bare. The leaves are long and narrow, and are a rich, green colour when new, but they darken in a few days.

FLOWERS: These six-petalled, fleshy flowers appear all around the year. Bees, butterflies, and birds love the flowers. They are pink and maroon, or can be multi-coloured with white or yellow on them. The blossoms have a strange, sweet perfume and hang from long, woody stalks, which grow out from the trunk or branches of the tree. When they fall, they lie in large numbers on the ground. If the petals are crushed, they immediately turn blue as the air reaches the crushed parts. In India, the flowers are used by worshippers for Shiva *puja*s, as the shape of the flower resembles a many-headed cobra protecting the inner part, i.e., the *lingam*.

FRUIT: The large fruits take eight to nine months to develop. They are round and grow to the size of a cannon-ball when ripe. Hard and brown on the outside, they contain sour-smelling, soft flesh, and many seeds. The cannon-balls, as they fall to the ground, break and turn bluish green and emit an unpleasant smell when exposed to the air.

USES: The tree is not of much practical use, but it is often planted in tropical countries for its beautiful flowers and odd appearance. The wood of this tree is of little value as it is of inferior quality. The shell of the fruit is sometimes used as utensils by natives of South America and in French Guiana, the pulp is fed to farm animals, and also eaten by the local monkeys. This is why one of its English names is Monkey Pots.

CASUARINA

Casuarinaceae
Casuarina equisetifolia R.& G.Forst.

WHERE IT GROWS: This tree grows wild from South India
through to Australia. It has been planted all along the coasts of India
as fire breaks and shelter belts. This practice is proving dangerous as
the tree has the ability to suck up moisture required by other plants and trees in the area. In this
way, it is similar to, and as dangerous as the Eucalyptus tree. It loves sunlight and sandy soil, and it
grows well on beaches and in exposed places. However, it also thrives inland and large numbers
often grow together. (Many trees of the same species growing together is called 'monoculture'.)

This tree is also called
She-oak, Australian Pine,
and Beefwood; it is known
as *vilayati saru* in Marathi;
jungli saru in Hindi; *chouk* in
Telugu, *savukku* in Tamil
and Malayalam; and *jhau*
in Bengali.

GENERAL DESCRIPTION: It is a tall, fast-growing, and short-lived evergreen tree with long, untidy branches. The bark is dark grey-brown, uneven, and peels off in patches. What looks like the grey-green, needle-shaped jointed leaves that appear to be segmented are, in fact, the branchlets. As these age, they become brown in colour and fall in heaps under the tree.

The name 'Casuarina' is believed to have come from a Latin word meaning cassowary, a bird from New Guinea. This is because the bird's feathers look like the spiky twigs of the tree. It is called Beefwood in Australia because under the bark, the wood is the same bright red colour as raw beef.

FLOWERS: In this species, male and female flowers are borne on different trees. The main flowering seasons are February and September to October. The female flowers start as buds covered with curly, dark red fur. This fur falls as the buds develop into small, round, woody cones. The male flower grows at the end of the needle-shaped branchlets in a fluffy, yellowish spike.

FRUIT: This is a wood-like cone which is green at first and then turns brown.

USES: In Europe, this tree is grown in hot houses and the branches are used for decoration. In tropical countries, it is used for preventing soil erosion, but as described earlier, it has environmental disadvantages that we are only now beginning to understand. In some places, the trees are trimmed and made into hardy hedges. This is done by cutting the stems and branches every time they grow too tall. In this way, they spread and form a thick hedge. These hedges are often used in 'topiary', which is the art of sculpting shapes of animals or birds out of the vegetation. This is a popular trend in parks and gardens all over India.

The wood is cheap and is often used as floor planks instead of the more expensive teak. It also makes good firewood. A dye from the bark is used to colour fishermen's nets and is also used as a medicine for stomach ailments and as a tonic.

IMPORTANT: If mangroves are planted in place of Casuarinas on the coast, erosion and damage caused by tsunamis and large waves can be reduced.

COCONUT PALM

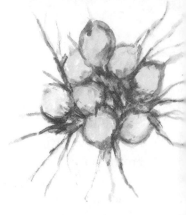

Arecaceae
Cocos nucifera L.

WHERE IT GROWS: The origins of this tree are still a mystery. It may have come from South America or the Pacific, with the nut being carried by sea tides and currents from place to place. It is, however, generally believed to have come from the Cocos Islands in the Indian Ocean. It has grown in India for many hundreds of years and is cultivated in all the hot, damp regions, particularly in the low-lying, sandy areas near the sea.

GENERAL DESCRIPTION: The palm takes about eight to ten years to mature (this means to grow to full size and bear fruit). Some cultivated palms, however, take as little as five years to produce nuts. By the time it is ten years old, this tree has a large number of feathery leaves. The tree is tall and medium- to fast-growing. It has a slender and usually straight trunk, ringed at intervals by the scars of fallen leaves.

It is called *nariyal* or *naral* in Marathi and Hindi; *dab* in Bengali; *tennai* in Tamil; and *tenga* in Malayalam.

The bottom or base of the trunk is swollen and surrounded by roots, and the bark is dark grey-brown and quite smooth. The wood is water-resistant and is often called 'porcupine wood' because of the attractive design on it.

LEAVES: These are large and feather-like, and they appear in a cluster at the top of the tree. Each leaf takes about a month to develop.

FLOWERS: Male and female flowers grow together on a branched stalk starting from the point where the leaves join the trunk. They are yellowish-orange and look like small catkins, which are spikes of soft, feathery flowers that hang from the twigs of trees like willow and birch. The male flowers are smaller and there are more of them, and they are scented.

FRUIT: About the size of a man's head, it has a green outer covering, which turns yellow and then brown when the nut is ripe. The string-like fibre covering the shell of the coconut protects the nut within.

Inside the nut is a liquid which sweetens as the nut ripens. This liquid then gradually becomes a thick white solid, which coats the inside of the nut and is called *copra*. Some trees can grow over a hundred nuts a year and this makes them very good crop trees.

USES: This is perhaps the most useful tree in India as nearly every part of it can be used. The string-like covering inside the outer case, called coir, is used to make rope, carpets, and mattresses. The ribs of the leaves are used for making kites and brooms. The wood is used for roof beams and house construction. From the sap or juice of the tree, palm wine or toddy is produced. The kernel or copra inside the nut is eaten in many different forms. When crushed, the copra gives coconut oil, from which candles, soap, margarine, and many other things are made. In Kerala, most of the food is cooked in coconut oil.

During the Second World War, injured soldiers urgently needing body fluids were given transfusions of coconut water by injecting the liquid from coconuts into the person's veins directly. This was because the water inside the nut was considered sterile, containing no bacteria or viruses.

Different types of coconut palms are being grown now. Not only do they produce nuts much sooner, but they are also shorter in height so that the nuts can be picked more easily.

COPPER POD

Caesalpinaceae/Fabaceae
Peltophorum ferrugineum
(Decne.) Benth.

WHERE IT GROWS: Originally from Sri Lanka and the islands near the coast of Malaysia, and North Australia, this tree is now planted in many tropical countries. It was planted in eastern India about 150 years ago, but only came to India's west coast recently, thus explaining the lack of vernacular names for this tree in Marathi or Malayalam.

GENERAL DESCRIPTION: It is a fairly fast-growing, tall tree with a straight trunk and smooth grey bark. It is a hardy tree and can live in most conditions without difficulty. From the trunk, many branches rise and spread outwards, giving a great deal of shade.

It is semi-deciduous and loses most of its leaves in winter but sometimes loses all its leaves for a few weeks before the new ones appear.

Its other English names are Rusty Shield Bearer, Yellow Gulmohur, or Yellow Flame tree; it is called *iya vakai* in Tamil and *kondachinta* in Telugu.

LEAVES: The large leaves are made up of small, smooth, leathery leaflets. These have delicate fur on the under-side. When they fall, they make the tree look ragged and bare. The new bright green leaves appear in early spring and darken in a matter of weeks.

FLOWERS: The main flowering seasons are from the end of February to April and then again from September to December. Some trees have a few flowers throughout the year. First, rust-red shoots appear and they are covered with soft hair and then buds of the same colour develop.

These open to show bright yellow, five-petalled, slightly scented flowers with wavy edges. They only last for a short time and the falling flowers leave a carpet underneath the tree. The tree has the strange habit of flowering in sections. One half of the tree will flower in one week and the other, in the next. In some cases, this depends on the direction the tree faces.

FRUIT: The name 'Rusty Shield Bearer' has probably been given to this tree because of the shape and colour of the seed-cases, which look like shields. These rust-red seed-cases remain on the tree for many months and later turn black in colour.

USES: It makes a good home for the lac insect. The wood of this tree is light and does not last very long, but it can be used for cabinet making. It is a popular tree in cities as its wide, feathery canopy is very beautiful and it is suitable for planting along roadsides.
Like the Australian Silver Oak, it is also used to provide shade for cocoa and coffee crops. The trees protect the growing plants from harsh sunlight, rain, and wind, and the crops thus protected often produce higher quality fruit. The bark of the tree produces a brown dye useful in the art of *batik*-making. It is also a useful garden tree as, unlike other trees, grass grows under it. Its name 'Peltophorum' is Greek for 'sheild-bearing', which describes the shape of the pods.

CORAL WOOD TREE

Mimosaceae
Adenanthera pavonina L.

WHERE IT GROWS: This tree has its home in mainland India, Myanmar, and the Andamans. It is commonly found growing wild in the Western Ghats and in parts of South India. It loves a moist and damp climate.

GENERAL DESCRIPTION: It is a very popular tree as it spreads out its branches to make a shady canopy and the seeds are very bright and pretty to look at.

It is a medium-sized to tall deciduous tree with a rough, dark grey-brown bark, and a slim, straight trunk. The branches usually start to grow high up on the trunk and then spread outwards. The younger branches have a smooth, red-brown bark.

In Bengali, it is called *rakta kambal*; in Hindi, *badigurochi*; *moti chanothi* in Gujarati; in Marathi, *thorligunj* or *ratangunj*; in Telugu, *bandi guruvendi*; and in Tamil, *aniakundumani*. In English, it is also known as Red Wood or the Red Bead tree.

LEAVES: Each single large leaf is made up of a lot of small leaflets, which grow on a stalk ending with one single leaflet. The leaflets are dark green in colour when fully grown but a fresh bright green when new. Where the leaves have fallen, horseshoe-shaped scar marks appear in the place where they grew. The leaflets are shed in the winter.

FLOWERS: The flowers appear from March to May. They are pale yellow and grow on spikes that look like small bottle brushes. These long stems grow from under the leaf stalks and at the end of the branchlets. The five petals fold back and the stamens stand out from the flower. The petals darken before the flower falls, but they are so tiny that they can hardly be noticed on the ground under the tree.

FRUIT: The seed pods, which are green when they first ripen, soon turn to dark red, and then brown. As they open, they peel back and twist to show a creamy, yellow, inner lining, and bright red, shining seeds. These seeds are all the same size and shape, and they fall to the ground in hundreds. They are called circassian seeds.

Although the seed pods appear at any time from the monsoon onwards, they often remain on the tree until after the next flowering season.

USES: Children love the seeds because they can be used as counters for games, beads for stringing, and in many other interesting ways.

Goldsmiths still use the seeds in some places for weighing gold as each seed weighs exactly the same. They can also be powdered and used to make medicines. The wood is sometimes used instead of red sandalwood for furniture or for wall panels.

The Coral Wood seed pods and seeds should never be mistaken for another similar pod and seeds called 'ratti' or *Abrus precatorius*. This is a slim, woody climber that grows all over the countryside and has highly poisonous seeds that are also used by goldsmiths for weighing gold. The seeds of ratti are usually bi-coloured red and black, and can be distinguished from the Coral Wood seed quite easily.

DRUMSTICK TREE

Moringaceae
Moringa oleifera L.

WHERE IT GROWS: This tree has its home in the Western Himalayas, but is now planted throughout India and in many other tropical countries. It produces an important commercial crop and is cultivated for its fruit in many Asian countries. It grows best in the sandy beds of rivers and streams.

GENERAL DESCRIPTION: It is a small, delicate, fast-growing, deciduous tree with a short life of approximately twenty-five years. It has a thick, grey, grooved bark, like cork, that peels off in patches. The wood is easily broken and is very soft.

It is also called the Horse-radish tree in English; *soanjna* in Hindi and Bengali; *shevga* in Marathi; *midhosaragavo* in Gujarati; *mulaga* in Telugu; and *murangakai* or *moringa* in Tamil and Malayalam.

LEAVES: The leaves are shed in December and January, and are clear green and rather feathery looking, with many leaflets on a long stalk.

FLOWERS: The long bunches of small, five-petalled flowers are greenish-white when in bud, creamy white when they open, and yellowish when withering. The flowers are honey-scented and attract many insects. The main flowering period is March to April, but many trees can be seen flowering from September onwards.

FRUIT: These ripen from April to June, although there are many trees that bear a few fruit all through the year. The fruits are long and stick-like, green in colour, and often deeply ridged. They hang downwards in clusters from the branches. The pods contain many seeds, each of which have three wings attached to them.

USES: Apart from the wood, which rots very fast, the tree is generally useful. The roots can be used for making horseradish sauce, which is a spicy sauce eaten with meat and which was particularly popular with the British who discovered this alternative to real horseradish during the days of British rule in India. The leaves and flowers are rich in vitamins, and can be made into curries. When crushed, the seeds give oil, which can be used in the preparation of perfume or for making the inner parts of watches run smoothly.

The fruits make a very popular vegetable and the pulp tastes rather like asparagus. One must be certain to scrape out the seeds and pulp when eating, however, and avoid the hard, fibrous outer husk. I have seen visitors to India nearly choke on drumsticks because they assume that every part of the fruit can be eaten! The pods are made into excellent pickles, especially in Tamil Nadu. The young branches are favourite fodder for cattle and camels. A coarse fibre is obtained from the bark, which used for making paper, mats, and string.

FLAME OF THE FOREST

Leguminoseae/Fabaceae
Butea monosperma (Lamk.)

WHERE IT GROWS: This tree is indigenous to India where it grows easily in drier regions all over the country. It is also abundant in Myanmar. The botanical name originates from the British Earl of Bute who named it.

GENERAL DESCRIPTION: Generally, this is a medium-sized, deciduous tree with a crooked trunk and twisted irregular branches. It is not a beautiful tree to have in one's garden and is at its ugliest in January when most of the leaves have fallen. One of the most written about and sacred trees in India, it is mentioned in the great epic stories of Ramayana and Mahabharata.

Also known as Parrot tree in English; it is known in Hindi as *palas, tesu, dhak, chalcha,* or *kankrei*; in Bengali, as *palas* or *palashi*; as *palas* in Marathi; in Tamil, as *porasum* or *porasu*; in Malayalam, as *muriki* or *shamata*; in Telugu, as *modugu*; and in Gujarati, as *khakda*.

LEAVES: These are tri-foliate and easy to spot. The tri-foliate nature of the leaves represents the sacred trilogy of Vishnu, Brahma, and Shiva for the Hindus. When fresh, the leaves are thick, velvety, and a delightful reddish-green in colour. In older leaves, silky down on the under-sides makes them appear silver. They generally start to open from March onwards after the short flowering season and lie alternately up the branches.

FLOWERS: From late January to March, black buds develop and burst into blossom. The tree is then transformed into clusters of scarlet-orange blossoms, all massed together at the ends of the branches. The flowers have five petals, of which one is longer than the others. This is like a hood to protect the other four and has given rise to the tree's other name, Parrot tree.

The flowers are pollinated by many species of birds and the tree, when it flowers, is a paradise for the bird watcher who has a clear view of the tree as it is leafless. There are trees that have yellow blooms, but I have personally not seen them.

FRUIT: The pods appear in May or June and are flat, oblong, and covered with velvety hair. They contain one seed and are greyish-green at first, turning yellow later. As the pods are very light when ripe, they are often blown far away from the tree. The seeds are kidney-shaped, brown, and flat.

USES: The tree is regarded as holy since it is considered to be sacred to Lord Brahma because of its tri-foliate leaves. Some tribal communities also say that it is sacred to the moon. The leaves are used in many Hindu ceremonies such as the rite of blessing female calves to ensure that they will produce plenty of milk later.

When the blossoms are collected and soaked in water, a beautiful yellow liquid is obtained. This liquid is an excellent lotion for the skin and is called *tesu*, or *kesu*. This yellow dye was used commonly as *gulal* (coloured powder) during Holi. Unfortunately, this tradition has been lost in many urban areas where toxic chemical dyes are used instead. These dyes permanently stain clothes and are also very bad for the skin.

It is an ancient belief, as with other sacred trees, that rags or flags should be placed on the tree to protect the giver from evil spirits. The leaves are sometimes used to make plates or other utensils and when they are young, they are fed to cows and buffaloes. The seeds are still used in some rural areas to relieve scorpion stings. The wood is very soft and not of much use, but it is used in villages for firewood.

GULMOHUR

Caesalpinaceae
Delonix regia (Boj.ex Hook.)

WHERE IT GROWS: Its home is Madagascar, earlier known as the Malagasy Republic. It came to India via Mauritius many years ago and is a very popular roadside tree because of its shape and the beauty of its flowers. It grows in many tropical conditions and is quite hardy, but it prefers dry areas near the sea.

GENERAL DESCRIPTION: It is a medium to large deciduous tree, very fast-growing, with a straight, slender trunk and smooth ash-grey bark. It has shallow roots which spread out around the tree and often do not allow other plants to grow nearby. The branches are brittle and easily broken, and they often get damaged in storms.

Often called Flamboyant, Royal Peacock Flower, Royal Gold Mohur, or Fire tree; the Hindi name is *gulmohur*; it is called *mayaram* in Tamil; *shima sankesula* in Telugu; and *gulmohr* in Marathi, Bengali, and Gujarati.

LEAVES: These are usually shed in autumn or winter, but they sometimes fall soon after the monsoon. Then the tree can remain bare in some parts of the counry until May or June. It has feathery-looking leaves like many other ornamental trees belonging to the pea family.

FLOWERS: These begin to bloom in the hot season from April onwards. At first, only a few open and later, they cover the whole tree in orange, bright red, and maroon. Four of the petals are orange-red and red. The fifth is larger with several shades of white and yellow and bands of red. The blossoms last only for a few days and open at night. Christians sometimes call this tree the Pentecost tree or Holy Ghost tree as it flowers at the time of Pentecost, fifty days after Easter.

FRUIT: The long, broad seed-cases, which are green in colour at first, gradually become hard and black. They remain on the tree for many months and they look very prominent after the leaves have fallen. The seeds are often soaked in hot water before planting to soften the outer case so that they germinate more easily.

USES: It is a tree that is largely grown for its beauty. The wood, which is white and soft, is used for making ornaments, and can be polished. The flowers and buds are used as herbs for flavouring food. Although this is an ornamental tree, some people do not like its bare appearance for many months of the year. The tree has a close relative which comes from Ethiopia and bears yellow blossoms. In French-speaking Madagascar, the tree is called 'fleur de paradis', meaning flower of paradise.

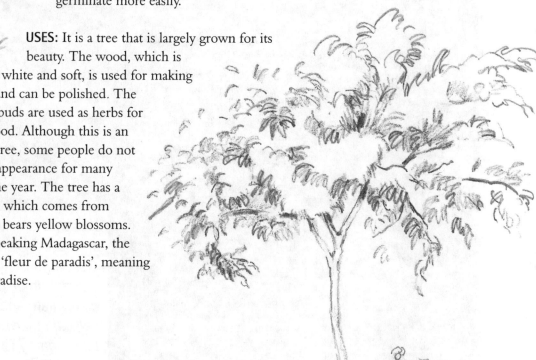

INDIAN CORAL TREE

Papilionaceae
Erythrina variegata L.

WHERE IT GROWS: This Indian tree is common in coastal forests and also grows wild in Burma (Myanmar), the Andamans, Java, and Polynesia. It grows well closer to the sea.

GENERAL DESCRIPTION: It is a very fast-growing, hardy, deciduous tree of medium height with a smooth, grey-green bark, which peels in patches. The trunk and branches are covered with sharp thorns that disappear as the tree gets older. The thorns protect the young tree from grazing

This tree is called mandara or *dadap* in Hindi; *palita mandar* in Bengali; *mandaram* or *pungu* in Malayalam and Telugu; *kaliyana murungai* or *mul murungai* in Tamil; and its Marathi name is *pangara*.

animals. The roots are shallow while the wood is brittle and easily broken. The tree of a related sub-species can grow up to 7000 feet high in the hills.

LEAVES: The leaves are made up of three, large, triangular-shaped leaflets, with the central one being the largest. These are bright green in colour and they fall in the winter, leaving the tree leafless until March or April. Young trees often keep their leaves throughout the year.

FLOWERS: These appear from early January and often continue growing up to March or April. In some parts of the country, and with some sub-species, the tree has a few blooms throughout the year. The bright scarlet flowers grow on spikes of blossoms, either alone or with others at the end of the smaller branches. Each spike has many blooms on it. The flowers have five petals, one of which is much larger than the others. The tree is especially beautiful in the spring when in full bloom. It is often planted with trees of the same family that have white or pink flowers, so that the two make a pleasing contrast of colours.

FRUIT: The seed-cases clearly carry the mark of the kidney-shaped, brown, red, or purple seeds inside them. The seed pods remain on the tree throughout the year, quickly turning black. They ripen from May to July.

USES: The soft wood is used for making small boats and for carving ornaments. The new leaves are used in curries. A red dye is obtained from boiling the petals. The tree is often used to provide shade for crops or for supporting grape creepers and pepper plants. It is grown as a hedge around gardens and a cut branch, when planted, can easily take root. The planting of these trees helps to nourish poor soil as the roots contain nitrogen-fixing bacteria that can convert nitrogen in the air into nitrates in the soil. These nitrates enrich the soil and help crops grow. Although the flowers have no scent, many birds and insects love the nectar, and are attracted by the bright colour of the flowers. The leaves are useful for feeding cattle, but the seeds are poisonous.

The Indian Coral tree is mentioned in the Mahabharata and there are many stories in Indian mythology that are associated with this tree due to its tri-foliate leaf structure.

INDIAN CORK TREE

Bignoniaceae
Millingtonia hortensis L.

WHERE IT GROWS: Originally from Burma and Malaysia, this is a tree that grows well all over India. It has been grown in this country for a very long time and grows wild in parts of central and South India. Its name '*Millingtonia*' is after a well-known English botanist from the eighteenth century.

GENERAL DESCRIPTION: It is usually planted because of its beauty and its fast growth. It has very shallow roots and does not allow many plants to grow underneath it. It is a tall, straight, evergreen tree with a dark yellowish-grey bark, which is rough and cracked. It has a few branches but these bend downwards and show off the lovely dark green leaves.

In English, it is called the Jasmine tree; in Hindi, *akas nim* or *nim chameli*; in Bengali, *mini chambeli* or *cork gach*; in Tamil, *karkku*; and in Telugu, *kavuku*.

LEAVES: The whole leaf is made up of small leaflets that grow in pairs up the leaf stem and end with one single leaflet. The leaves are shed gradually during early spring, from January to March, so that the tree is never bare.

FLOWERS: It flowers from April to June and then again in the autumn, from September to December, which is the main flowering season. The waxy-looking flowers are shining white in colour and some have patches of pink inside them. They have a long, tube-like lower end to which four petals are attached. One of these has a deep split in it so that it looks as if there are five petals. The flowers grow in bunches on stems at the end of branchlets. When they fall, the ground becomes carpeted in white. At night, the blossoms are more sweetly scented than during the day. They attract moths who, because of their long proboscis or tongue, have the ability to reach into the flower and sip the nectar. While doing so, the moths also pick up pollen and transport it to another bloom or tree.

FRUIT: This tree does not produce seeds very easily. They grow inside long, thin seed-cases with pointed ends and each thin seed has a delicate wing.

USES: The bark of the tree can easily be pulled off in chunks and it is often used to make a rather poor kind of cork. The wood is very soft and breaks easily. Therefore, it often snaps off in a strong wind and is not very suitable for planting along roadsides or near buildings. The wood is used for furniture and is very pale yellowish-white in colour. The best quality cork comes from parts of Spain and is infinitely superior to the wood of the Indian Cork tree. When branches break off, a new stem grows in its place very fast, making the tree look untidy.

INDIAN LABURNUM

Caesalpinaceae
Cassia fistula L.

WHERE IT GROWS: Native to India, this tree grows wild all over the country up to about 4000 feet. It also grows well in almost all the tropical and sub-tropical areas of the world. Many countries have claimed the Laburnum as indigenous. Greece and Egypt even use ancient literature to support their claim. It is thought, however, that India is its rightful home. It is described by many as being one of the loveliest flowering trees, infinitely more beautiful than the European Laburnum.

Its other English names are Golden Shower, Pudding Pipe tree, and Golden Rain; it is sometimes referred to by its Hindi and Bengali name, *amaltas*; in Bengali, it is also called *sondal*; in Marathi, it is called *bahava*; in Gujarati, *garmalo*; in Malayalam, *konna* or *kritamalam*; in Telugu, *rela*; and in Tamil, *sarak-konne* or *tiru kontai*.

GENERAL DESCRIPTION: This tree is small to medium in height, upright, slow-growing, and deciduous. The branches are slender and drooping, and the young bark is smooth and greenish-grey, becoming brown and rough as the tree ages. It is a tree that often grows by itself, as most of the seeds are consumed by animals and carried away before they can germinate.

LEAVES: Between March and May, the old leaves fall and the tree is nearly leafless. The new leaves start to develop in early June. The leaves are fairly large and oval in shape, lying opposite each other in pairs of three to eight, on short stems. They are an attractive copper colour when new, with soft down on the under-side and they remain drooping and folded until mature. Once fully grown, they are a smooth, dark green above with silvery hair below and they are quite distinct from any other form of Cassia, which has feathery leaves and long, arched branches.

FLOWERS: As the weather gets hotter, the blossoms appear as long, drooping sprays of clear yellow flowers from twelve to eighteen inches long. They cover the tree in a mass of fragrant golden blooms that catch the eye. The individual flowers sit in a green cup or calyx, and the five petals are unequal in size and spoon-shaped. Inside each bloom are ten stamens, three of which are much longer than the others and curve upwards in a stately fashion. The blossoms come in many shades of yellow and there is even an almost white variety.

FRUIT: This is a long pod, about two to three feet in length, which is green and soft at first, becoming brown later, and eventually, black and hard. These pods hang in profusion from the tree and take almost a year to ripen. They are very conspicuous in the spring when the tree is bare.

Each pod contains as many as a hundred shining, flat, brown, oval seeds in a sweet pulp. The name Pudding Pipe tree or Monkey Sticks refers to the pipe-like shape of the pods but really do not do the tree's beauty any justice.

USES: This tree has many uses. The roots, bark, seeds, leaves, and dark pulp of the fruit are used as a purgative to make one vomit, and in lesser doses, are an effective laxative for constipation. Poultices of the leaves are said to relieve chilblains, which are itchy painful sores on the hands and feet caused by extreme cold. The poultices are also good as a cure for rheumatism. Strangely, the purgative properties that are very strong in humans do not seem to affect monkeys, bears, or other animals that devour the sweetish, black fruit pulp and show no ill-effects. The timber is hard and long-lasting, and is used for agricultural tools, posts, and carts. It also has a pretty grain and is used to make fine pieces of small furniture. The Santals, who are the largest adivasi community in India, use the flowers of this tree as food.

JAVA PLUM

Myrtaceae
Syzigium cuminii

WHERE IT GROWS: This tree originated in India, Myanmar, Sri Lanka, and Malaysia, and it grows all over the country except in very dry, sandy areas. It also grows well in Australia.

It is very common in Maharashtra where forests of this tree can be seen. A tree of the plains, it does not grow well at altitudes over 4500 feet, where other related species replace it.

GENERAL DESCRIPTION: It is tall and fast-growing after the first two years. An evergreen tree, it can grow in many conditions and is fairly hardy. It has a pale or dark grey-brown bark, which is rough and thick, and peels off in patches. The branches start high up the trunk and make a wide tent of leaves. The trunk is often crooked.

In English, it is also known as the Black Plum, Indian Blackberry, or Indian Allspice; in Hindi, it is called *jamun* or *jambul*; in Bengali, *kala jam*; in Marathi, *jambhul* or *jaman*; in Tamil, *naval* or *naaga*; and in Telugu, *neredu*.

LEAVES: These grow facing each other on the stems, some are thin and lance-like, some others are oval, and others are almost round. If the leaves are held up to the light, tiny transparent spots can be seen all over them. These spots can be seen on many of the leaves of the trees of the Myrtle family, of which the Jamun is a member. The trees lose their leaves in early spring, usually during January or February. When the leaves are crushed, they emit a strong smell. They are often fed to tasar silk worms.

FLOWERS: These are greenish-white in colour and very tiny. They are bunched together in groups on stalks that grow out under the leaves. They are very sweetly perfumed and they open from March to May.

FRUIT: These ripen in June or July and sometimes, a little earlier. They are oval in shape and dark purple when ripe. They vary in size but can grow to the size of a pigeon's egg. They are very juicy and sweet when ripe, but the juice is astringent and dries the mouth. Ponies love the *jamun* fruit and can often be seen with their mouths covered in pale purple froth. (When my daughter was a little girl, she was frightened that the ponies were suffering from rabies.)

Many people do not like to plant this tree in their gardens as the fruits make a mess when they fall. A related species that grows at higher altitudes has smaller fruit.

USES: This tree is considered sacred by the Hindus, especially for Lord Krishna, so it is often planted near temples. It is also planted for the shade it gives on the sides of roads in cities. Liquor, vinegar, jellies, and preserves are made from the juice of the fruit, and many parts of the tree are used in local medicine. The wood is hard and can be used for boat building and general construction purposes.

MANGROVES

GENERAL DESCRIPTION: Mangroves are one of the most productive ecosystems that exist in the world today and also one of the least understood. They are disappearing at an incredibly fast rate, even faster than tropical rain forests. This is due to human carelessness and natural disasters such as the tsunami that struck our coasts with such ferocity in December 2004.

What is so special about mangroves and where do they grow? Mangrove forests used to cover three quarters of our tropical and sub-tropical coastlines. Now less than 50 per cent remain and much of this is degraded or spoilt. Most people know very little, if anything, about the benefits of these vital plants.

The roots of mangrove plants stabilize the sand and mud in which they grow. This prevents erosion and also helps to build up the shore line and stop it from receding or disappearing into the sea. Mangroves also act as barriers against heavy tides, tsunamis, and cyclonic winds that often batter the coast. They do this by absorbing the heavy flow of water or wind, thus preventing damage to the shore. When the tsunami hit the Indian coast, most of the damage was in areas where mangroves had been deliberately removed. The Sunderbans in Bengal and some of the more unspoilt islands in the Andamans were relatively untouched.

The other important environmental factor is that mangroves protect marine life and allow fish, crocodiles, prawns, and other animals to breed in safety. The mangrove trees grow in salty or brackish water. (Brackish water is water that is a mixture of salt and fresh water.)

To quote from the Mangrove Action Project, 'If there are no mangrove forests, then the sea will have no meaning. It is like having a tree with no roots, for it is mangroves that are the roots of the sea.'

The Mangroves

Rhizophoraceae
Rhizophora mucronata Lam.

WHERE IT GROWS: They grow in the Sunderbans in Bengal and along the coasts and backwaters of India.

GENERAL DESCRIPTION: They are small evergreen trees supported by enormous roots that hang from the stem and branches, and bury themselves in the mud. The branches are purple in colour.

In Bengali, the plant is called *bhora*; in Marathi, *kandal*; and in Tamil and Telugu, *upoomoma*.

LEAVES: They are elliptical, thick, and bright green in colour. The under-side of the leaf is pale and covered with tiny, dark spots.

FLOWERS: The flowers are white, fleshy, and hairy blossoms sitting in a pale yellow calyx.

FRUIT: The Mangrove's fruits are long, dark brown, hanging pods.

USES: A fermented wine can be made from the juice of the fruits.

There are many members of the Rhizophoraceae family, all of which are mangroves, and all of which are interesting to botanists because of their particular adaptations or characteristics that allow them to live in salt water.

Some, like the one above, have aerial or brace roots that lift the tree up above the high tide mark. Others have special breathing roots that grow up out of the water against gravity. They look quite strange when one sees them for the first time. Some mangroves have seeds that germinate while still inside the fruit and then later drop down into the water and grow, or get washed away by the tide and germinate elsewhere.

The Blinding Tree

Euphorbiaceae
Excoecaria agallocha L.

WHERE IT GROWS: This tree grows from Africa, across the Asian subcontinent, to Japan, Australia, and the Pacific. It usually grows above the high water mark and can thrive in both stony and muddy ground.

GENERAL DESCRIPTION: It is small bushy tree with rough grey bark that looks like cork.

LEAVES: These are oval with pointed tips. They are pinkish when young and turn bright red when they are ready to drop off.

The other English names of this tree are Milky Mangrove or Blind your Eye; in Malay, it is called *buta buta*; in Bangladesh, it is called *gewa*; it is called *tillai* in Tamil and *thilla* in Telugu.

FLOWERS: The tiny yellowish-green flowers grow in spikes. Male and female flowers bloom on separate trees.

FRUIT: The fruits are small and round, and are grouped in clusters. They float in water and explode when ripe to disperse or throw out the seeds. The seeds also have an air capsule to help them float.

USES: The milky sap irritates skin and can cause temporary blindness, thus explaining the common English name. The plant is used to treat sores and smoke from burning the bark is used to treat leprosy. Clinical trials have shown that the plant may contain chemicals that can treat HIV-AIDS and that it may have anti-bacterial, and anti-viral properties as well.

Indian Mangrove

Avicenniaceae
Avicennia officinalis L.

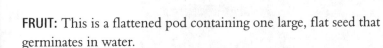

WHERE IT GROWS: This tree grows on the coasts of southern Asia and Australia.

GENERAL DESCRIPTION: This is small evergreen tree with thin, brownish-grey bark that becomes rough and either black or yellowish-green as the tree matures. This is a tree that grows in or near salty or brackish water and sends up narrow, erect, breathing roots called pneumatophores. These roots are the tree's method of breathing and make the tree look as if it is balancing on stilts.

> In Hindi, it is called *baen*; in Bengali, *bina*; in Malayalam, *orei*; in Tamil, *madai pattai*, or *upattha*; and in Telugu, *mada* or *nallamada*.

LEAVES: The leaves are oblong, thick, and leathery with the edges slightly rolled under. They are shining green above but covered in grey-green hairs and tiny black spots on the under-side.

FLOWERS: The blossoms are bell-shaped, like small tubes, with four lobe-like petals. They are yellow or yellow-brown in colour, turning orange as they mature. They are grouped at the end of the twigs and each bloom sits in a hairy, five-lobed cup. Their scent is very unpleasant.

FRUIT: This is a flattened pod containing one large, flat seed that germinates in water.

USES: The tree is used a great deal in folk medicine. The unripe seeds are used as poultices for boils and the fruit is also used as a treatment for tumours. In some places, a resin from the bark is used as contraceptive. The Javanese eat the bitter fruit and seeds after cooking it in a particular way. The wood has an attractive grain and is useful in cabinet making.

MAST TREE OR CEMETERY TREE

Annonaceae
Polyalthia longifolia (Sonn.)

WHERE IT GROWS: This tree is native to the drier parts of Sri Lanka and the south of India, but it is now planted in many parts of the country along roads, by rivers, and around buildings and temples. It grows well in poor soil and its trunk is strong enough to withstand monsoon winds. Still, it does not grow very well close to the sea.

GENERAL DESCRIPTION: It is a tall, slim, majestic, evergreen tree that grows rather slowly. Its bark is smooth and dark greyish-brown.

Also called Indian Fir in English; in Hindi, it is called *ashoka* or *debdaru*; also *debdaru* in Bengali; *asopalav* in Gujarati; *nara* or *nanameamidi* in Telugu; and *assothi* or *mettilingam* in Tamil. The tree should not be confused with the other Ashoka tree, which is a short, spreading tree with a heavy canopy, dark green leaves, and orange-scented flowers.

Although the tree is generally tall and slim, another form of this tree spreads outwards with weeping branches that hang down. The leaves, flowers, and fruit of both the forms, however, are similar.

LEAVES: The leaves are dark green and shiny on both sides. This tree probably once grew in very wet areas as the leaves droop downwards and rain drops pour off the shiny surfaces of the leaves, keeping the tree dry.

FLOWERS: The six-petalled, yellowish-green, star-like blossoms appear in February, March, and April—the main flowering months. The blossoms grow in bunches and produce a great deal of fruit.

FRUIT: They are small, egg-shaped, and green in colour before ripening. They generally ripen in the period between July and September. Each fruit contains one seed. The ripe fruits are purple but they are difficult to see as they remain hidden in the leaves. Flying foxes and other bats, monkeys, and all birds love the fruit. They leave the ground covered with the tree's seeds that they discard during their feasting.

USES: This tree is planted for its beauty and is held sacred by Hindus who make wreaths and garlands of the leaves. Earlier, the straight trunks were used to make masts for sailing ships, but now, the wood is used only for making matches, pencil boxes, scaffolding, barrels, and drums.

In Malaysia, the tree is often grown around cemeteries and is associated with funerals. There is a similar tradition in Europe, but there, Cedar trees are planted instead—both trees seem to give the same sad feeling.

MIMOSA-LEAVED JACARANDA

Bignoniaceae
Jacaranda mimosifolia D. Don

WHERE IT GROWS: This tree is a native of Brazil and Argentina, but over the last hundred years, it has been introduced into most tropical and sub-tropical regions. I have seen some spectacular specimens in Antananarivo, the capital of Malagasy Republic. There are also some huge trees in the mountains of South India, where they seem to thrive even at attitudes of 6000 feet or more. This is strange as the tree is supposed to be sensitive to frost, but this does not seem to have any effect on its ability to grow well.

This tree has very few local names that I could discover, but in Hindi, it is called *nil-gulmohur* or *jamuni badal*; in Bengali, *nilkantha*; in Tamil Nadu, it is simply called *jacaranda*; the other English names are Palisander or Green Ebony; in Malay, it is called *jambul merak*; in parts of Sri Lanka, it is called *gualanolay*.

GENERAL DESCRIPTION: It is a medium to large semi-deciduous tree with a wide, spreading crown once the tree matures. When young, the fast-growing trunk and branches are long and leggy, with a few frond-like leaves at the top. They are therefore rather unsightly and I tend to cut the newly-grown stems so that they can grow into an attractive canopy. If one does not do this, one has to wait for about seventy years to get the same effect! The trunk is covered in a smooth grey or black bark.

LEAVES: The elegant foliage is fern-like and symmetrical. The tiny leaflets, which are arranged in pairs of twenty or more, are pale green when young and darken as they mature. They lie opposite each other and have pointed tips. About eight to sixteen sets of leaflets lie along a stalk and this gives the feathery appearance that is the characteristic of the tree.

FLOWERS: The real beauty of the tree are its blossoms, which are bluish-mauve, tubular, and which sit in a five-lobed calyx. Up to ninety or more sprays of drooping blooms cluster at the end of the twigs. Each flower is like a bell and has lips that curve outwards. It is a glorious sight to see a mature tree in full bloom, especially as the new leaves are hardly visible. The flowering season is from March to May and sometimes, later in the year.

FRUIT: There are not many fruits and that ripen between June and November. They are round, flat, wavy-edged disks containing may tiny seeds with papery wings. Green when young, they turn brown as they ripen. The seeds fly in the wind, often long distances from the parent tree.

USES: It is grown as an ornamental tree in many parts of India. The bark and leaves are used extensively in Colombia for medicinal purposes. In India, an infusion of the bark, obtained when soaked in water, is used to treat ulcers. Part of the bark and leaves are a good treatment for syphilis and chest complaints. The wood has an attractive grain and is quite hard. It can be used for making cabinets, small furniture, and tool handles.

MOHWA

Sapotaceae
Madhuca indica J.F. Gmel.

WHERE IT GROWS: This tree is native to central India and extends north to the Himalayas and south to Sri Lanka. It also grows well in Myanmar. It prefers to grow in areas away from the coast. Its scientific name means 'honey' in Sanskrit.

In English, also called the Butter tree; the Gond tribals of Bastar call the tree *gondi* or *idu*; in Oriya, it is *mohulo*; in Hindi it is known as *mahua* or *moha*; in Bengali and Gujarati, it is called *mahwa*; in Tamil, it is called *kaata illupai*; in Telugu, *ippa* or *ippi*; and in Malayalam, it is called *poonam*.

GENERAL DESCRIPTION: It is a large deciduous tree with a straight trunk and a thick, dark grey-black bark that is cracked vertically and which is very wrinkled.

LEAVES: The tree is bare in the winter and new leaves only start to appear in March or April. The young leaves are bright rust or crimson in colour and leathery in texture. As they grow older, they change to a bright green and are bunched together on long stalks at the end of grey branchlets. If the stalks are broken, they produce a white, rubbery, liquid sap. The leaves appear at the same time or shortly after the flowers.

FLOWERS: The blossoms open at night and by the morning, fall to the ground. They are fleshy and cream coloured. Before the petals open, they are completely covered in a furry, rust-red calyx. The flowers hang in bunches at or near the ends of the branchlets. They are tubular in shape and very sweetly scented, although the scent is slightly musty and not always pleasant.

FRUIT: The fruits are egg-shaped, juicy, green berries containing one to four shining brown seeds. These ripen from June to August.

USES: The value of this tree, which is seldom cut, comes from its fruit and its sugary-sweet blossoms. Many adivasis in central India rely on the flowers as an addition to their diet. In the early morning, during the flowering season, tribals can be seen sweeping up the blooms, which are then eaten raw or cooked. They can also be fermented and distilled into a potent but unpleasant tasting liquor. Often, the flowers are stored as they can last indefinitely and then sugar can be extracted from them. It is said that one tree can produce over 160 kilograms of flowers in one season! Animals, particularly deer and monkeys, are very fond of the flowers, which ferment in their stomachs and cause intoxication, so they behave in a completely drunk manner! Parakeets, mynahs, and other birds are also known to be affected by the fermented juice of the blooms.

The seeds are crushed for oil, which can be used instead of ghee. The seeds can also be made into soap and candles. The outer fruit is eaten as a vegetable. The white milky sap is used to treat rheumatism and the oil is a cure for constipation. Any wood that is cut is used for furniture, carts, and agricultural implements.

In Bihar, the tree is worshipped during marriages and the Gond tribals use the tree not only for food and medicine, but also as a totem against witchcraft. The folklore surrounding this tree is extensive.

MOUNTAIN EBONY

Caesalpinaceae
Bauhinia variegata L.

WHERE IT GROWS: There are many species of the *Bauhinia* family in the tropical world, of which Mountain Ebony and several others are indigenous to India. The other most common tree in the family is *Bauhinia purpuria* or the Camel's Foot or Geranium tree. The species *Bauhinia* is named after twin brothers, Jan and Caspar Bauhin, early botanists who lived in Switzerland in the sixteenth century. The Mountain Ebony, in common with other members of its family, is easily recognized by the strange shape of its leaves.

In Hindi, the tree is called *kachnar*, *padrian*, *gurial*, or *kandan*; in Bengali, it is called *raktakanchan* or *kanchan*; in Marathi, *kanchan*; the Tamil name is *segappumanthari*; in Malayalam, it is known as *kovidaram*; in Telugu, *mandarii*, *deva*, or *kanchan-amu*; and in Malay, *kupu-kupu*.

GENERAL DESCRIPTION: It is a small to medium-sized deciduous or semi-deciduous tree, which is generally smaller when cultivated in gardens outside India. It has irregular branches and the crown looks crooked when viewed from a distance. The bark is dark brown and fairly smooth.

LEAVES: The most distinctive feature of this tree is its dark green, thick, and rather dull-looking leaves. The whole leaf is kidney shaped, but divided into two lobes that are heart shaped at the base and the leaf has a large cleft or split at the top. The veins on the leaves spread out like a fan. The leaves grow alternately up the twigs on short stalks. The tree is virtually leafless from February to April and remains so until after the flowers start to bloom. Some trees retain a few leaves all through the year.

FLOWERS: The flowering season is from February to April. The blossoms are large and delicately scented. They grow in short sprays of three or more blooms, either from the ends of the branches or in the angle between the leaves and the main twig. The calyx or cup in which the petals sit is greyish-brown with silvery hair and it has five ribs which split as the flower opens. The flower also has five petals, one of which is larger than the others. Each petal has delicate veins on it and often, the colour at the base is deeper than the colour above. The blooms can be magenta, mauve, or pink, with splashes of crimson. There is also a white variety which has bright yellow markings. At a glance, the blossoms look like orchids.

FRUIT: The pods are shiny, long, flat, and narrow, often more than a foot in length. They contain five to sixteen seeds. Once the brown fruit is ripe, it bursts open, throwing the seeds at some distance from the tree. The case of the pod then twists and curls as the seeds fly. The fruit ripens in May or June.

USES: *Bidis*, Indian cigarettes, are often wrapped in the leaves of the Mountain Ebony. The tree also produces a useful gum called 'cherry gum' which is sugar-free and can be used for making sweets. Oil is extracted from the seeds and the bark is used for tanning leather. A juice from the roots is reported to be an antidote to snake bites and is also used to control obesity. The young leaves and flower buds are eaten or pickled, and also make good fodder for cattle. Sometimes, the tree is grown as shade for coffee bushes. *Bauhinia* is held sacred to both Buddhists and Hindus. The name Mountain Ebony describes the reddish-brown, heavy, and hard timber.

MUDILLA

Barringtoniacae
Barringtonia speciosa (Forst.)

WHERE IT GROWS: Originally from the Malagasy Republic and the East Pacific, this tree grows well in tropical Asia and Australia. It loves sandy beaches and thrives in most coastal regions where there is a lot of salt in the soil. Even though it is a fairly common and beautiful tree in cities, not much has been written about it. If you live on the coast, it is a tree to look out for. It is closely related to the mangrove species that protect coastal areas.

The Singhalese name for this tree is *mudilla*; in Malay, the tree is known as *putat laut*, *butun*, *butong*, and *pertun*; in English, its names are Beach Barringtonia and Fish Killer tree; in Hindi, it is called *hijjal* or *injar*; in Bengali, *hinjolo*; in Marathi, *piwar*; in Tamil, *adampa* or *kadapa*; in Telugu, *kurpa*; and in Malayalam, *piwar*.

GENERAL DESCRIPTION: It is a medium-sized, fairly slow-growing, evergreen tree. It has a dark brown or almost black bark, which is rough and uneven. The younger and smaller stems are smoother. The straight trunk has branches quite high up and these fan out to give it a beautiful green crown.

LEAVES: Dark green, thick, and shiny, the leaves have clearly marked veins. They grow in groups at the ends of the branches. The new leaves are paler in colour and a brighter green.

FLOWERS: The blossoms grow in bunches on stems at the end of the branchlets. They open at night and are visited by bats and moths who pollinate them. First, large, round, green buds appear. These open and expose the large flower, which is made up of four fleshy white petals that flatten out and contain a mass of long, pink-tipped stamen. The flowering season is generally September to October, but in many places, a few blossoms can be seen throughout the year. In this case, it is moths and bats that carry out the process of pollination. When they fly onto another flower, they cause the pollen they are carrying to fall onto the pistil so that the fruit can be formed.

FRUIT: The fruit consists of a large, green, triangular seed case, which turns brown after ripening and has a lid to let out the seeds. As this seed case floats, it is often carried by the sea to other parts of the coast or even to other countries. Inland rivers also carry the fruit to other places.

USES: The wood is soft and does not last long. It is sometimes used to provide fishermen's floats. (These are pieces of light wood or cork attached to the fishing nets to mark their position in the water.) The bark and fruits are poisonous, and can be made into powder used to paralyse fish so that they can be easily caught. As the cooking of the seed destroys the poison, the fruit is eaten in Indo-China. A Hawaiian government law states that any preparations of *Barringtonia speciosa/asiatica* 'are unlawful to possess or use on, in, or near state waters for the purpose of taking aquatic life' (Law number 188 – 23). This law covers most of the substances used to kill fish, including pesticides and other toxic material. As Indian law is not specific, people can use any toxic material to kill fish and aquatic life in reserves and national parks. This license to use toxins causes immense harm to the environment, bringing damage to all fresh water and marine life. Since *Barringtonia* only paralyses certain fish and does not affect other fresh or marine organisms, it would is be a safer and more environmentally-friendly alternative to industrially-produced toxins.

NEEM TREE

Meliaceae
Azadirachta indica A. Juss.

WHERE IT GROWS: This tree originated in India and upper Myanmar but it now grows in Sri Lanka and Malaysia as well. It is a hardy tree and grows in many conditions, preferring dry areas where it can grow to an enormous size. It does not grow in the hills and it survives only in frost-free zones.

GENERAL DESCRIPTION: It is a medium to large-sized, evergreen tree and is fairly fast growing with a rough, pale grey-brown bark. The trunk rises to quite a height before the branching starts. In drier areas, this tree becomes semi-deciduous.

LEAVES: New leaves appear throughout the year, but they form mainly in March and April to replace the old leaves that have fallen.

Sometimes called the Margosa tree; *nim* or *neem* in Hindi; *nimgach* in Bengali; *limba*, *timba* or *nimbay* in Gujarati and Marathi; *vepa* in Telugu, Tamil, and Malayalam.

The leaves are fresh green, shiny, and sometimes rust-tinted when new. They crowd together near the ends of the branches. Since it is thought that the leaves absorb carbon dioxide faster than other trees during photosynthesis, more oxygen is released by the Neem tree. For this reason, it is a healthy tree to grow in gardens or near houses.

FLOWERS: The tiny, pale cream coloured, and star-shaped, five-petalled flowers hang under the leaves and are difficult to see. The tree flowers from March to May and again during the rains. Their strong perfume attracts many insects.

FRUIT: The small, almost transparent berries, green at first, become yellow as they ripen. Birds and bees love their sweet-tasting juice. After a rain, the fruits give off a strong and unpleasant smell.

USES: These trees are often planted along the roads to give shade, particularly in North India. The tree has many medicinal uses. The most famous product of the tree is the oil obtained from the seed called margosa oil, which is used for treating skin diseases and leprosy. The leaves are used for treating boils. The wood is heavy and drives away insects. It is also used as a bio-pesticide and has the ability to regulate nitrogen in the soil. I use neem cakes profusely in my garden to kill pests.

The twigs are used in place of tooth brushes and tooth paste, particularly amongst tribals and the poor in the countryside. In North India, the trees grow so well that there is often a sturdy branch onto which a swing can be attached.

The Neem tree symbolizes health. The story goes that many years ago, a man was going on a journey which would take him away from home for many weeks. His wife advised him to sleep under a Tamarind tree on the outward journey and a Neem tree while travelling homewards. After a few nights of sleeping under the acidic Tamarind, the man became sick and was forced to return home, but he was completely healthy after a few nights under the Neem.

Every part of this tree has its uses as the active ingredient in the tree is azadirachtin. It is a pity that India does not utilize the tree more efficiently.

PALMYRA PALM

Arecaceae
Borassus flabellifer L.

WHERE IT GROWS: This palm is a native of India and it grows extensively from the Ganges river basin in the north to Cape Comorin in the south. It also grows in Myanmar and Sri Lanka. The botanical name, *flabellifer*, comes from the fan-like shape of the leaves, which are quite different from those of many other palms.

GENERAL DESCRIPTION: This tree can grow to 100 feet in height and older trees often have a slight thickening in the middle of the trunk. Because the trees are tall and slim, they often bend in the direction of the prevailing wind. When young, the trunk of the palm is covered with old leaves or spiked with the jagged stalks of fallen leaves, but as the tree grows older, the trunk becomes

The other common English names are Fan or Toddy palm; it is called *tal, talgachh,* and *tarkajhar* in Hindi, Bengali and Marathi; *panai* in Tamil; *karimpana* in Malayalam; *tadu* or *tadi* in Telugu.

smoother and dark grey or black in colour. At the base of the tree is a mass of root fibres that twist and writhe like snakes just above the ground.

FLOWERS: The trees are either male or female and have different male and female flower spikes. The female flowers are larger than the male and often look like small fruits. They are scattered sparingly. There are many more of the smaller male blossoms. Both flowers are enclosed in long, branched sheaths or covers. The male flowers are covered by bracts that seem to be sunk into the branches. A bract is a type of modified leaf that comes between the flower and the leaf stalk. The flowers can be seen twice a year, from March to April and again later in the year.

FRUIT: The female tree produces a large cluster of small, oval fruits, the outside of which are leathery and tough. Under this covering is a layer of fibre. The fruits start off green, then they turn orange and finally turn black when ripe. Once the outer covering has been cleaned, which is a job for an expert, a small, transparent, slippery, jelly-like fruit covered in a thin orange skin is obtained. Inside this fruit is a clear liquid, which tastes very much like coconut water. In Tamil, this fruit is called *nangu* and in Marathi, it is called *targola*.

Young Palmyra Palm

USES: Tamil poets have described the palmyra palm as having over eight hundred uses as practically all parts of the tree are useful. The wood is made into rafters and pillars or split to make water channels in rural cottage construction. The leaves are used for thatching, mats, fans, baskets, country umbrellas, sandals, plates, and many other purposes.

As evening approaches in many parts of the country, men climb the tall trunks using only a loop of rope to support them. They make a V-shaped cut in the flowering branches of the male and female trees which have earlier been crushed. From this cut comes a thick, sticky liquid which pours into a cup placed under the slash. If the pots are first lined with lime powder, the juice becomes a thin liquid, and when drunk early in the morning, is a refreshing drink called 'neera'. Much, however, is allowed to ferment and becomes slightly alcoholic. This is called 'toddy' and can be distilled into 'arrack', which is a strong spirit. The rest of the liquid is used for palm jaggery or sugar, and for making excellent vinegar. The orange outer skin of the fruit is very bitter but it is regarded as a sure cure for dysentery.

PEEPAL TREE

Moraceae
Ficus religiosa L.

WHERE IT GROWS: The Peepal tree comes from the Himalayan foothills and Myanmar, but is now planted all over the country, and also in Burma. It is another member of the fig family.

GENERAL DESCRIPTION: It is the largest of the trees belonging to the fig family and because it is regarded as sacred both to the Hindus and Buddhists, it is often planted near temples and holy places. It is supposed to be one of the longest-living trees. There is one in Sri Lanka that is said to be over two thousand years old!

This tall tree is nearly deciduous and grows very fast. The bark is light grey and peels off in patches.

Instead of the hanging aerial roots that grow from the Banyan's branches, the Peepal trees have roots that are attached to the trunk so that they look as if they are pillars supporting it.

Sometimes called the Bo or Bodhi tree; in Hindi, it is known as *pipal* or *peepal*; in Bengali, as *asvattha* or *ashathwa*; as *jari pipro* in Gujarati; *pimpal* or *ashvathi* in Marathi; *ashvatthamu* in Telugu; and *arasu* in Tamil.

The tree needs a lot of space and the soil must be deep enough to let the roots grow a long way down. There are many tales about this tree. Lord Buddha is said to have found enlightenment while meditating under a Peepal or Bo tree at Bodh Gaya in Bihar.

LEAVES: The long, pointed leaf tips help to drain water off the leaves and dry the tree after rainfall. Leaves are shed from March to April and in some areas, in the autumn months. When the new leaves appear, they are often pink and darken to copper, and then to green. The leaves are rarely at rest and flutter even at the slightest breeze.

FRUIT: Like the Banyan, the flowers of the Peepal are hidden within the figs. These figs ripen in May and June, but they are also found at other times of the year in different areas. The fig wasp is a visitor to these fruits as well as to the Banyan and other fig trees.

As with the Banyan, the fruits are eaten by birds or bats, and the seeds are dropped to start new growth in all sorts of places. It is a common sight to see a Peepal tree growing from a gutter or even from the wall of a house. When this happens, the tiny plant gets all its food from the air and water, and is called 'epiphytic'.

USES: Nearly every part of the tree can be used as medicine. From the bark, a reddish dye can be extracted. The leaves are used to feed camels and elephants, and the tree is often the home of the lac insect. As with all the fig trees, many birds and bats love to eat the fruit, and in times of little food, villagers eat them as well.

The wood lasts well in water and is sometimes used for building small boats. As the tree is revered, it is often covered in offerings or cloth, and it is also considered to be haunted. A branch of the famous Bo tree in Bodh Gaya was brought to Sri Lanka about 1500 years ago and is planted in the ancient city of Anuradhapura. Very often, devout Hindus marry a Banyan tree off to another tree such as Neem. This is considered to be a sacred duty.

PERSIAN LILAC

Meliaceae
Melia azedarach L.

WHERE IT GROWS: Native to the lower Himalayas, including Kashmir and Baluchistan, it is also found in upper Myanmar. This attractive tree has been planted for ornamental purposes in many countries and for centuries, it has been revered in Iran, Malaysia, China, Sri Lanka, and India, where it is often planted near places of worship. It has now become naturalized in Southern Europe and America.

The other English names for this tree are Chinaberry tree, Bead tree, and Indian Lilac; in Hindi, it is called *bakain* or *dake*; in Bengali, *bakajam*, *wilayati nim*, or *gora nim*; in Gujarati, it is known as *bakan*; in Malayalam, as *karim vembu* or *vaymbu*; in Tamil, *malei vembu*; and in Telugu, it is called *yerri vepa*.

GENERAL DESCRIPTION: A slim, upright, fast-growing, deciduous tree of medium height. It is short-lived and as it grows older, it acquires a dark grey-brown bark that has long fissures down the trunk. The branches are wide and tend to break easily in the wind. Since its roots are shallow, it is not suitable for urban areas.

LEAVES: The leaflets are bright green and lie opposite each other on a long stalk that connects with a major leaf stem. They are narrow and pointed with serrated edges, and spray out at the end of the branches in groups of five or six. The leaves fall in winter and open again in March. As the leaves are similar to the Neem tree's, the two trees are often confused with each other.

FLOWERS: The blossoms open before the leaves are fully developed. They are beautiful sprays of tiny lilac blooms with a delicate fragrance. Each tiny flower has five or six, narrow, mauve or white petals that open to show the deep purple stamens forming a tube inside the blossom.

FRUIT: In July or August, round, green fruits appear, each one containing five seeds. These berries remain on the tree in bunches, turning bright orange first and then black, and often remaining on the tree for many months before falling to the ground to germinate. Each of the seeds has a natural hole through it. The fleshy part of the fruit smells very unpleasant when ripe and birds tend to avoid these trees. I have, however, seen bulbuls feeding on them occasionally and they do not seem to be affected by the poison in the fruit.

USES: In Southern Europe, the tree is very popular as the seeds are used to make rosaries. The fruit is said to have narcotic properties and when eaten in large quantities, it can be poisonous. The berries, bark, and leaves contain insect-repellant properties. Flea-powder and bio-insecticides are still made out of this tree. Over a hundred years ago, inflammable oil made from the fruit of this tree was used in lamps.

Both the roots and leaves of the tree can be consumed to remove intestinal worms and the seeds are sometimes used as a cure for rheumatism. There are many parts of the tree used in medicines and like the Neem, it is a very useful tree. The wood is tough, moderately hard, durable, and is never attacked by termites. It has an attractive grain and is used to make small pieces of furniture, toys, and other articles. In parts of South America, it is grown as a shade tree in coffee plantations.

PONGAM

Papilionaceae
Pongamia glabia Vent.

WHERE IT GROWS: This tree is found growing from the Seychelles to India, the Pacific islands, and in Australia. It grows all over India and is planted for its beauty by the sides of roads and in gardens. It is very hardy and can thrive in any weather. It grows especially well on coasts and near streams and rivers, but it also grows well at fairly high altitudes. This tree grows so commonly along the coast of Maharashtra that a peninsula not very far from Mumbai is called Karanja, after the tree. Its species name comes from Tamil (*Pongamia*), and 'glabia' means smooth and hairless.

It is called Indian Beech in English; *karanj* or *karanja* in Hindi, Bengali, Marathi, and Gujarati; *pongam* in Tamil; and *pungu* in Telugu.

GENERAL DESCRIPTION: It is a fast-growing, medium-sized, nearly evergreen tree with a short trunk and spreading crown of branches. It has a smooth, dark grey-brown bark.

LEAVES: These are shed at irregular intervals in winter and spring. When new leaves appear, they are an eye-catching bright green, but they soon become a dull green. It is common to see brown patches caused by caterpillars on the leaves. In fact, many insects and fungi attack this tree. It is called Indian Beech because during February and March, when the new leaves unfold, it looks very much like the European Beech tree.

FLOWERS: The stems of pale purple or white flowers appear soon after the new leaves appear around February, March, and April, carpeting the ground with their blossoms, which soon wither and dry. In some places, the flowering season is very irregular, with some trees bearing blossoms in summer or autumn, or even twice a year.

FRUIT: The woody seed-cases are fairly small and a patchy dull-yellow or dark brown in colour when ripe. The seed case is reminiscent of a raft or rubber dinghy. They ripen just before the new leaves appear but they remain holding the seeds until the seed-case decays. The seed is red in colour. Wherever the seed-case falls, the seeds germinate easily and in some places they have to be removed; otherwise, too many trees would grow. As they often flourish along river banks, the floating seed-cases are carried a long way from their parent tree.

USES: The wood is pale yellow in colour and does not last long. It is sometimes used for making cart wheels and other implements. When cut, the branches are fed to cattle or ploughed into the land to enrich the soil. The seeds produce oil which is useful for treating skin diseases and for burning in lamps and making soap. The oil has a strong unpleasant smell. The juice of the roots has antiseptic properties.

QUEEN'S FLOWER

Lythraceae
Lagerstroemia speciosa (L.) Pers.

WHERE IT GROWS: This tree belongs to Assam, Burma, the Western Ghats, South India, and Sri Lanka, but it also grows in Myanmar, and up to China and Australia. It is very common on river banks and in marshy areas.

GENERAL DESCRIPTION: It is a fast-growing, deciduous tree, which can grow tall in wet areas. It remains small or of medium height in dry or unsuitable places. Hardy in rich soil, it can often

In English, its other names are Pride of India, or Crepe Myrtle; in Marathi, it is known as *taman*, or *mota bondara*; in Hindi it is *arjuna* or *azhar*; in Bengali, *ajar*; in Tamil, *kadali*; and in Telugu, *vara gogu.*

flower only two years after planting. It has a smooth, light grey bark, which peels in patches. The branches grow fairly high up the trunk and produce a crown of leaves. In Kerala, there are forests of these trees which look very beautiful in the flowering season.

LEAVES: Leaves fall gradually from January onwards and turn a reddish brown before falling. The tree is never completely bare. The leaves are large, oval-shaped, and rough with veins clearly visible. The upper surface is a dull green colour, but it is paler underneath. The new leaves grow in pairs along the branches. In early April, the new leaves appear, covering the tree in a very short time. They are a clear pale green.

FLOWERS: These appear in April and remain until June, often flowering again in July and August. They grow on long spikes at the end of the branches and are very showy, six- or seven-petalled, purple flowers turning paler as they fade. The petals have a crushed and crumpled look, like tissue paper. The extraordinarily long flowering season of this tree makes it particularly popular as an avenue tree.

FRUIT: These are woody, round seed-cases with five or six openings. They are green in colour at first but later change to brown and black. The seeds are pale brown and smooth with a stiff wing.

USES: The wood of this tree is very hard and is rated as valuable timber for construction purposes. It is also used in boat-building. Many other parts of the tree are used medicinally or for producing tannin. All parts of the plant are useful in the treatment of diabetes and in the Andamans, the fruits are used to cure mouth ulcers. The roots are astringent and the seeds are used for their narcotic effects. The tree is one of the loveliest trees grown in rural India and has now also become a popular urban tree.

RAIN TREE

Mimosaceae
Enterolobium saman Prain.

WHERE IT GROWS: This tree comes from tropical America and the West Indies, and was first introduced in Sri Lanka and then in India. It is a tropical tree and cannot stand cold climates. It prefers moist areas.

GENERAL DESCRIPTION: It is a tall, fast-growing, nearly deciduous tree, which has the ability to keep its shape even with strong winds blowing towards it from one direction. It has shallow roots

Otherwise known as *saman* or *guango* in Marathi; in Hindi and Gujarati, it is called *vilayti siris*; in Tamil, it is known as *thungamoonji*; in Bengali it is called *belati siris;* in Malayalam, *plavu;* and the other English names of this tree are Cow Tamarind and Monkey Pod.

and a dark grey bark. The trunk can become very thick and its canopy can become so wide that it is not a tree to be planted in a small garden. In cities, the spread of the crown keeps many roads cool and shady.

LEAVES: The oval, dull green leaflets can change their position according to weather conditions. In full sunlight, they lie flat and open, but at night and in cloudy weather, they swivel and fold, so that they lie sideways. The leaves are shed in winter, but the tree is never completely bare.

FLOWERS: The flowering periods are March to May and again in late autumn. In many cases, the tree flowers all through the year. The flowers are grouped together in bunches that look like balls of fluffy cotton-wool. The flowers are pink and they show up because of their grouping. A very similar tree called *Albizia lebbek* has many fuzzy, green and white, scented flowers that resemble the pink flowering heads of the Rain tree. The fruit of this tree are long, yellow pods that clatter in the wind. The tree is similar in shape to the Rain tree.

FRUIT: These are flat, fleshy, brown seed-cases that contain a sweet pulp, which surrounds the seeds.

USES: The tree is supposed to have been brought to Sri Lanka to make good fuel for steam trains. However, it was found that it did not burn well. Still, it is commonly planted on the sides of roads and gardens, and is used to protect coffee and cocoa crops from the sun. The fruits are very popular with squirrels and are often fed to horses and cattle. They are said to increase the quantity of milk produced by cows. The seeds are not digested but are swallowed and then passed out of the animal's body. In times of scarcity, the leaves are used to feed animals.

This tree is also the home of the lac insect, but the lac is not of a very good quality. It is called the Rain tree in Malaysia too where people believe that when the leaves fold, rain will follow. The tree also seems to have a habit of spraying moisture onto the ground, which gives it its name. But this moisture is actually produced by insects and not by the tree itself.

SILK COTTON TREE

Bombacaceae
Bombax malabaricum DC.

WHERE IT GROWS: The Silk Cotton tree is believed to have come from the area stretching from Malaysia to North Australia, but it grows wild in most parts of India except in very hot, very cold, or dry areas. It loves damp, sandy soil and grows well in coastal regions, especially in the Konkan.

GENERAL DESCRIPTION: This tree lives very long. There are Silk Cottons that are believed to be almost a thousand years old. It is a tall, deciduous tree, very easy to pick out in a forest or jungle in the spring when it is bare of leaves and covered with flowers. It has a straight, thick, upright trunk, which is often wider at the base to support the heavy branches above. The branches grow out horizontally. Most trees have branches that first grow out sideways and then curve upwards. This feature makes the Badam, True Kapok , and Silk Cotton trees very easy to recognize. The bark is

In Maharashtra, this tree is called *kate-savari*, *lal savaar*, or *semla*; in Hindi, it is known as *simal* or *shimbal*; in Bengali, *ragtasimal*; in Tamil and Malayalam, *illavu* or *elavum*; and it is called *salmali* in Sanskrit.

smooth silver grey and covered with hard, sharp prickles. As the tree gets older, the bark roughens and cracks, and many of the prickles disappear.

LEAVES: The whole leaf is made up of close to seven small leaflets, which are oval or lance-shaped, and have a long leaf stalk that joins the smaller branches. They are bright green in colour and leathery in texture. When they are on the tree from April or May until the winter months, they provide a great deal of shade.

FLOWERS: From January to March, the flowers appear and they are easy to see on the leafless tree. The flowers are usually bright red but they can also be pink or yellow, or white. They have five, large, curved petals that contain many long stamens. The male part of the flower contains pollen. The stamens are grouped in bunches. The flower sits in a fleshy cup calyx and is divided into three sections that are silky and green in colour. Nearly all flowers have a calyx, although they are all different in shape, colour, and size.

FRUIT: The green seed-cases start to ripen in April and are oval, but they soon thicken and turn brown and hard when ripe. They split and the soft silky cotton inside is blown away by the wind. This cotton contains the oval, black seeds from which new trees germinate and grow.

USES: There are many folk stories about this tree. One name given to it is Parrot's Despair, probably because the seed-cases look so delicious, but only contain cotton. This cotton has many uses as it is good for filling cushions and quilts, and is even sent to other countries. Unfortunately, like the Yellow Silk Cotton tree, the cotton, although very comfortable to use, has soft fibres which soon break and this means the additional filling of cushions and pillows. The gum, which is obtained from the trunk, is called *mocha-ras* and is used as medicine and for book binding. The flowers, along with their green calyx, are eaten as vegetables. They are also loved by animals, especially deer, who make a meal of the fallen petals. When the tree is in flower, it is always covered with birds and insects, especially drongos, mynahs, and sunbirds, who love the nectar. The holes in the trunks of old trees are often used as nests by parakeets. The tree is considered sacred to Lord Shiva and to Buddha who, according to legend, was born under this tree.

TAMARIND

Caesalpinaceae
Tamarindus indica L.

WHERE IT GROWS: Originally said to have come from Central Africa, there is no record of its arrival in India. It grows well in all tropical countries and in Southern Europe, and is very adaptable to most climates although it does not like frost or high altitudes.

GENERAL DESCRIPTION: Tamarind is a slow-growing tree, nearly evergreen in moist areas and nearly deciduous in dry areas. It is a large, tall tree with a short trunk and a crown of leaves.

It has very long, deep roots and its bark is thick, almost black, cracked, and uneven. It can live for a very long time. As the tree is very acidic, many small ground plants do not grow underneath it.

In Hindi, it is called *imli*; it is called *tentul* in Bengali; *chincha* in Marathi; *amli* in Gujarati; and *puli* in Tamil and Malayalam; the Persian name *tamar-i-hind* means 'Indian Date'.

LEAVES: These fall and are replaced almost immediately, normally in April, May, and June in dry areas. The young leaflets, which together form the whole leaf, are small and thick. They are bright green when new but soon darken to a dull dusty green.

FLOWERS: Blossoms are hardly noticed but are very pretty and scented. They are pink, purplish cream, or yellow veined, and hang in loose bunches from the ends of the branches from April to June, and often later in the year.

FRUIT: They ripen from February to April and there are usually a great many of them. It is said that there is less fruit on the trees in North India than in the south. That is perhaps the reason why most South Indian food contains tamarind while North Indian food uses tomato as a souring agent. The seed-cases differ in shape. Some are long, some are curved, others are flat and small. They are green at first and covered with brownish fur. Later, they turn dark brown or reddish black and brittle. They become stiff and are easily broken. The pulp inside the seed-cases surrounds the seeds and is sour.

USES: The tree is often planted in parks and along avenues. Its wood has a regular grain and is very hard. The wood is often used for making charcoal, wooden hammers, and furniture. It is also used for making the pin on the potter's wheel. The fruit is a favourite for making curries and pickles, especially in South India.

The seeds are ground to make flour, which is sometimes used for making chapattis. The flour is also used as starch for stiffening cloth and has other uses in the textile industry. The seeds are used commercially as they contain pectin, which is used for making jams and jellies.

In Gwalior, there is a Tamarind tree over the tomb of Tansen, the musician from Akbar's court. By tradition, singers eat the leaves of the tree to improve their voice. The tree is nearly always found where there are people and villages. This means that it has always been planted for use and is not a wild tree.

TEAK TREE

Verbenaceae
Tectona grandis Linn.f

WHERE IT GROWS: This tree comes from central and southern India, Myanmar, Thailand, and Java. It is fairly adaptable but grows best inland and loves a warm, moist, tropical climate with plenty of rain. In Kerala, teak trees were planted in the 1840s and have now grown to an enormous size, but elsewhere, the timber has been harvested so often that most trees are fairly small.

Teak trees also need plenty of space, light, and good soil. They are often found near rivers where they can grow enormously tall.

GENERAL DESCRIPTION: The tree is a fairly slow-growing, deciduous tree and it takes from sixty to eighty years to mature. The bark is

It is also called Indian Oak. It is called *saagun* or *sewan* in Hindi; *saka* in Sanskrit; *sag* in Marathi; *saguna* in Bengali; *saga* in Gujarati; and *tekku* in Tamil and Malayalam. This tree is often refered to as Rangoon or Burma Teak.

ash coloured or brownish-grey, and scaly. Teak trees cannot be cut down without government permission. For this reason, some people are afraid to grow them in their gardens, although with permission, people are allowed to use the timber. When travelling in the countryside from June to September, you should look for this beautiful flowering tree.

LEAVES: Large and strong, the leaves grow in pairs and are shed from November to January in dry areas. In moist areas, they often remain on the tree until March. The tree remains bare during the hot weather and the new leaves start growing early in the monsoon season. In some areas, certain insects eat the leaves, leaving the tree looking wild and untidy.

FLOWERS: Pyramids of tiny, white, scented blooms appear from June to September. They stand above the leaves in a pyramid shape and make the tree look as if it is covered in tall triangular candles. The flowering stems remain on the tree long after the flowers have died.

FRUIT: They look like tiny, crushed, green Chinese lanterns, and contain a furry nut. The seeds inside take a long time to develop and are white and kidney-shaped.

USES: The tree is famous for its wood, although its quality depends on where the tree grows. Some areas in India produce fine wood, but Burma Teak is considered to be the best. The wood contains a resin which preserves it and stops insects or termites from eating it, so that even the poorer quality wood has many uses. The leaves are used as plates and for covering roofs. When the tender leaves are scratched, a red colour appears and this is used for dyeing cloth. The red colour that appears is a sure sign of genuine teak. The ash of the burnt wood is supposed to be very good for the eyes. Many tribals worship plants and trees as totems for protection against evil, and the Teak is one such tree for the Bhil tribals.

TEMPLE TREE

Apocynaceae
Plumeria rubra Linn.

WHERE IT GROWS: This tree comes from Mexico, Guatemala, and tropical American countries, but it is now commonly grown all over India and in Sri Lanka, where it has found a new home.

GENERAL DESCRIPTION: It is a small or medium-sized deciduous tree with a fleshy, crooked stem and branches, and is fast-growing and hardy in most conditions. The bark is grey and smooth, and if any piece of the stem breaks off and is planted, it will often take root and grow in a short time. If any part of the tree is damaged, a white, sticky sap oozes out. The tree grows well up to about 4500 feet. Above that level, it may survive, but it will not flourish. In fact, in my garden, I have two plants that have been there for over four years and still have only three leaves and stand only a foot high.

Its other popular names are *son champa*, Frangipani, and Pagoda tree; in Maharashtra, it is called *khera champa, son chafta*, or *pandhra chapha*; in Hindi, it is called *chameli* or *lal gulechin*; in Bengali, it is called *dalama phula*; in Gujarati, it is known as *rhada champo*; in Marathi, *khair champa*; and in Tamil, as *segappu atali*.

LEAVES: These are smooth, dark green, oblong, and pointed at both ends on short leaf stalks. They grow alternately up the stems, one after another. The leaves grow in large numbers from winter to the rainy season, and then the tree remains nearly bare for the rest of the year, except for young trees, which often keep their leaves throughout the year.

FLOWERS: The lovely flowers are at their best from March to May, but often the tree will have some flowers all around the year. They grow in bunches on an upright stem at the end of the branchlets. They are white with bright yellow centres and sometimes the petals have little patches of pink. The petals, which are five in number, are waxy and twisted, and have a very sweet scent. When they drop to the ground, they remain unblemished and can be used for decoration. There are several other closely related species that have yellow (lutea) or white (alba) flowers.

FRUIT: In India, these do not tend to develop regularly, but I have seen several fruit-bearing trees in the Mumbai area and also in Tamil Nadu. The seed-cases are first like two green sticks and then become oval, brownish, and woody, sticking out like horns from a long stalk. They contain thin, flat, oval seeds with delicate wings.

USES: This tree is often planted in temple gardens and for this reason, it is called the Temple tree. The flowers are used for pujas. Many parts of the tree are used for medicines, for example, the white sap is used for treating rheumatism. This white sap is also a substitute for rubber.

Buddhists, Christians, and Muslims all consider the tree to symbolize immortality. This is because even when the tree is uprooted, it still can produce leaves and flowers. It has also been called the Tree of Life. The white soft wood is sometimes used for making drums.

TRUE KAPOK TREE

Bombacaceae
Ceiba pentandra (L.) Gaertn.

WHERE IT GROWS: This tree is native to South and Central America, the Andamans, Malaysia, and other tropical countries. It is also grown commercially in many other places such as India, Sri Lanka, the Philippines, and Cambodia. It is the national tree of Nicaragua. Although not a beautiful tree, it is easily recognized and stands out in a forest or grove.

GENERAL DESCRIPTION: It is a medium-sized, deciduous tree that can grow very tall if left untouched. When young, the stem and branches are covered with many sharp prickles. This is a defense mechanism against marauding animals. The base of the trunk is buttressed or thickened to support the weight of the tree above. From the straight trunk the branches grow out at right angles in a whorled or twisted fashion, giving the tree its distinctive, if strange, look.

Also known as Singapore Kapok or White Silk Cotton tree in English; in Hindi, it is *safed simal* or *katan*; in Bengali, *swet simal*; in Malayalam, *seemapooli*; in Tamil, *tella burunga*; and in Telugu, *illa vum*.

LEAVES: The leaves fall in winter but start to appear again in December. The leaflets grow on short stalks and are joined at the base onto a long leaf stem or petiole. There are usually five to eight leaflets that are lance-shaped and look like fingers. The arrangement of these leaflets is called digitate, meaning exactly like fingers. The entire leaf consists of these joined leaflets on the leaf stem. They are bright green in colour.

FLOWERS: The blossoms appear between January and March, and are small and dirty white, or light pink in colour. They have a slightly milky smell which is not unpleasant. The flowers hang in clusters near the end of branches and appear before the leaves are fully developed. Each bloom has five fleshy petals sitting in a green, bell-shaped cup. As the flower opens, the five stamens appear as a tube joined at the base and fan outwards so that they are clearly visible. Compared to other members of the Silk Cotton family, the blooms of this tree are insignificant and cannot compare with the brilliance and size of the flowers of the Yellow Silk Cotton tree described in this book.

FRUIT: Each pod is long, oval-shaped, and pointed at each end. It has five compartments, each of which contains many round, brownish, or black seeds. Each seed is attached to and covered by soft silky hair that grows on the inside edges of the fruit.

USES: Although the tree is often planted along roads, its main product is the springy, insect proof, elastic fibres in the pod. These are used for stuffing cushions, mattresses, chairs, pillows, and quilts. Since the fibres are covered in a thin, waxy coating, they are water repellant and unsuitable for spinning. But this is one of its most valuable attributes. The fibres are used to fill life jackets and lifebuoys. During the Second World War, sailors used to refer affectionately to their life jackets as kapoks. Silk cotton also floats better than cork. Many cottage industries in India rely on this tree for stuffing and prefer natural fibres to the more expensive man-made materials. The fibres also provide excellent insulating material against heat and sound, even though man-made fibres have, in many cases, replaced natural material. The oil from the seed is used as food, as a lubricant, and as an ingredient in the manufacture of soap. Some medicines are also extracted from the tree. The wood is used for matchboxes but is otherwise light, soft, and pretty useless.

TULIP TREE

Bignoniaceae
Spathodea campanulata Beauv.

WHERE IT GROWS: This tree originally came from tropical Africa and was then brought to Sri Lanka and India. It grows well in most tropical countries, particularly in drier areas. In moist areas, it does not grow to the same large size as it does elsewhere.

GENERAL DESCRIPTION: This is a tree that can be damaged by high winds as the wood is soft and the branches are easily broken. Proverbs in Africa use the name of this tree to describe people who are weak-willed and cowardly.

The other English names are Bell Flambeau, Fireball tree, Gabon Tulip tree, Scarlet Bell tree, Fountain tree, Syringe tree, Squirt tree, and it is sometimes wrongly called Flame tree. It is called *rugtoora* in Hindi; *patade* in Telugu; *patadi* in Tamil; and *tulip brikshya* in Bengali.

It is a fast-growing, tall, semi-deciduous tree in dry areas and in moist areas, it is of medium height and semi-evergreen. It is a narrow, slender tree with upright main branches and a straight trunk. The bark is light grey-green and rough. New trees are usually grown by cutting a piece of a leafy stem and planting it. After some time, this cut section produces its own roots and grows into a new tree.

LEAVES: This tree has deep green, oval leaves that fall during February and March. They are coarse and have deeply marked veins.

FLOWERS: These also appear in February and March with the leaves, but they often go on flowering into the monsoon. There is a second flowering season from October to December. First, velvety brown buds appear in huge bunches at the end of the branches, followed by blooms of orange, bright red, and dark red, which are beautiful to look at.

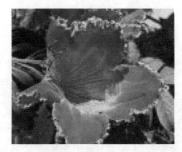

When the flower buds are squeezed, a jet of water comes out, and boys use them as water pistols.

FRUIT: It is not reported to produce fruit very often in Maharashtra, although in hotter and drier areas, this problem does not arise. Even when the seeds do appear, they apparently have some difficulty in germinating. They are lance-shaped, smooth, woody, and dark brown, and contain papery, winged seeds.

USES: Often planted for its beauty, it makes a showy avenue tree and is used in South India as a useful shade tree for coffee or tea. In Africa, the fruits are boiled and a poisonous liquid is obtained, which hunters use to kill their prey. The soft white wood does not burn and is seldom used, except on the sides of blacksmiths' bellows or for African drums. When cut, the wood gives off a strong garlic smell. The wood is fire-resistant and is used as a boundary for crop fields. It is also used in Africa and Haiti to make witch doctors' wands as it is thought to have magical properties.

WILD DATE PALM

Arecaceae
Phoenix sylvestris L. Arecacea

WHERE IT GROWS: This is a native tree that grows best on the Coromandel Coast and in West Bengal, although it can be seen fairly often and virtually everywhere in India, from the plains up to altitudes of about 4500 feet. It loves grasslands and open wasteland, and is very hardy, needing very little soil, and often seems happy in the most inhospitable and rocky areas.

GENERAL DESCRIPTION: The palm is slow-growing and small compared to other palms, rarely reaching more than 30 feet. It has a scaly trunk, broad at the base, which narrows as it rises up. The scales are the scars of the fallen leaves, which remain as reminders of past fronds.

It is known in Hindi as *khajur*, *khaju*, *shindi*, or *shindad*; in Bengali, as *kejur*; in Tamil, as *icham-pallam*; in Malayalam, as *kattinta*; and in Telugu, as *pedda-ita*.

I have a *Phoenix Loreiri*, which resembles the Wild Date Palm in my garden, but this species hardly ever has a trunk more than two metres high and often, much less.

LEAVES: The feathery, greyish-green fronds spring from the base of the trunk in an arch and are very attractive. They divide into rigid leaflets ending in spines that are sharp. The leaflets spread out far from the crown or trunk and shade plants growing beneath.

FRUIT: These are tiny, sweet dates that are green when raw and turn dark brown or almost black when ripe. They grow on a long, yellowish-orange stalk, which stands out above the fronds. Animals, particularly palm squirrels, and many birds fight over the sweet fruit. People also eat the fruits, although they are are usually eaten in times of hardship when other fruits are not available.

USES: Like many palms, most parts of the tree can be utilized in some form or another. Toddy or palm wine, a popular drink, is made from the sap tapped from the trunk of the tree. If this toddy is distilled, it makes a lethal spirit called *paria-arak*, which is consumed on the Coromandel Coast. When the sap is fresh and then boiled, it produces a delicious, soft, black sugar, which is highly prized and quite different in taste from other jaggery. It is available in large quantities during the spring in Bengal and people travel long distances to buy it. In the south, the leaves are cut to make brooms and in many other areas, country baskets and mats are woven from the young fronds. Rope is made from the twisted leaf stalks.In some places, the fruit is chewed with betel nut as paan. Considered a utilitarian tree, it is not held sacred by any religious group.

The Date Palm that we are more familiar with and which produces large fruits is called *Phoenix dactylifera*. This palm comes from West Asia and North Africa and is known to be one of the first trees to have been cultivated for food. Records of this palm date back over 5000 years. This tree is much taller than its smaller Indian relative.

WILD MANGO TREE

A nacardiaceae
Spondias pinnata (Linn.f.)

WHERE IT GROWS: The original home of this tree probably was in the Assam-Burma region, but for thousands of years, the mango has grown all over India from the Himalayas to the south.

GENERAL DESCRIPTION: This hardy tree is semi-evergreen. It grows fairly fast and can grow to an enormous size in the wild. The trees that are specially grown are never as large and are known only for their fruit. There is no record of the age of mango trees, but some definitely live for hundreds of years.

In North India, it is known as *am* or *amb*; in Hindi it is *amara* or *amba*; in Marathi also it is *amba*; the Tamil name is *marmaram*; the Gujarati name is *amri*; and the Malayalam name is *amram*; in English it is also known as Hog-plum or Traveller's Delight.

They have a huge crown of branches that act as a tent. In the countryside, the tree is used as a meeting place for villagers. The bark is dark and thick like cork and peels off in patches.

LEAVES: The leathery leaves are lance-shaped, smooth, and are grouped at the end of the branches. When crushed, they give off a strong smell of mango. The new leaves are a very attractive pink, which protects them from intense sunlight.

FLOWERS: The main flowering season is from December onwards, but in many parts of the country, the trees flower at other times of the year as well. Pyramids of tiny, pale, four or five-petalled blossoms grow together on a branching stalk. The blooms have a strong perfume that attracts insects and bats. Only a few of the many flowers are perfect and capable of producing fruit.

According to legend, the daughter of the Sun God was being chased by a witch and to escape, she jumped into a lake and turned into a lotus. A passing king fell in love with the blossom and the witch was so angry that she burnt the lotus to ashes. From these ashes grew the mango tree.

FRUIT: It is said that there are almost a thousand varieties of mangoes and though most bear fruit in the summer, there are many that bear fruit throughout the year. Not all the fruits are good to eat. In fact, many wild mangoes taste bitter and unpleasant (the flavour is often compared to turpentine). The fruits differ in size, shape, texture, and colour and they hang on long stalks at the end of the branches.

USES: The wood of the Mango tree is used for making furniture, boats, and dugouts, but it is soft and lasts for a very short time, although it lasts longer under water. The fruit contains high amounts of Vitamin C. The tree is supposed to be a manifestation of Prajapati, the lord of all creatures, and is sacred to the Hindus. Mango leaves are therefore used to decorate houses on festive occasions.

Mangifera indica is a tree closely related to the Wild Mango tree that produces excellent quality fruit, but this does depend on soil and many other factors. Some of the mango varieties that are most sought after are Alphonso, Neelam, Langra, Deshari, and Banganpalli, although there are many more varieties that different regions specialize in.

YELLOW SILK COTTON

Cochlospermaceae
Cochlospernum religiosum (L.) Alston

WHERE IT GROWS: Its name *Cochlospernum* is taken from the Greek word for spirally twisted or snail-shaped. The tree is a native of India, Myanmar, and Malaysia, and can grow in the rockiest ground. It thrives in dry, deciduous forests and is particularly spectacular near mountain roads or ghats in South India. It grows well up to about 3000 feet. Often planted near temples, the beautiful flowers are picked for daily worship.

GENERAL DESCRIPTION: It is small to medium-sized deciduous tree, untidily branched with a slim, upright trunk. The bark is ash-coloured, thick, and smooth, but as the tree ages, furrows appear,

Also known as Torchwood, Cotton Shellseed or Buttercup tree in English; in Hindi, as *kumbi*, *galgal*, or *gooloc*; also *galgal* in Marathi; in Bengali, *golgol*; in Gujarati, as *kadachogund*; in Malayalam, as *pakutakka*; in Tamil, as *kannigaram*, *kattilavu*, or *tannaku*; and in Telegu, as *adaviburaga* or *gungu*.

from which a sticky gum oozes. In winter, when the tree is leafless, the crooked branches look rather unattractive.

LEAVES: These are clustered towards the end of the branches and fall in winter so that the tree is almost completely bare. The leaves develop after the flowers in March or early April and are palmate. They are dark green above and greyish below. The young leaves have soft down on the under-side, which disappears as the leaves mature. Each leaf grows on a long, thick petiole.

FLOWERS: The large, beautiful, and eye-catching flowers are bright yellow, comparable to the colour of the European buttercup, thus explaining one of it English names. The buds appear in small clusters at the end of the branches. The flower is enclosed in five silky sepals, which are the segments of the calyx or cup in which the petals sit. The blossoms are also five-petalled and look glorious on the stony hills or plains where they tend to flourish. They stand out against the darker trunk and branches. The blooms appear in late February or March and only last for a short while.

FRUIT: The fruit is a round, dark brown capsule as large as a goose egg. It has five lobes and is full of kidney-shaped seeds cushioned in soft silk. The fruits ripen in May or June when the pods burst open and release the seeds. These are borne by the wind many miles away from the parent tree.

USES: The tree is planted for ornamental purposes and Buddhists often offer the flowers in temples. The silk is used for stuffing pillows and cushions, and is excellent for splints and as additional padding in bandages. Unfortunately, the fibres of the silk tend to be delicate, so stuffing needs to be re-done frequently. When I first came to India, I used this silk cotton a great deal, but soon found it rather expensive and impractical. The gum from the trunk is used in villages by shoemakers and for book binding. It can also be given as a sedative and cure for coughs, and it has a sweetish taste.

In rural areas, the wood makes for good torches because of the inflammable gum but it is otherwise soft and useless. Medicinally, the dried flowers and leaves can be used as a stimulant. In some parts of India, the wood soaked in water and later mixed with flour is consumed as food.

WHAT TO LOOK FOR ON TREES

1. The size of the tree and whether the branches grow out from the trunk high up or lower down, near the ground. Also whether the branches start at right angles from the trunk or curve upwards.
2. Whether it is a shady tree, a tidy or untidy tree, or a prickly tree. Is it very tall or very short? Do the prickles grow out above animal grazing level or stop at a certain height?
3. Notice the colour of the bark and whether it is smooth or rough, peeling or uneven.
4. See if the roots grow above the ground or are twisted into the trunk of the tree.
5. Look for any flowers on the tree and if there are flowers, look to see what colour and shape they are. Note the number of petals and whether the stamens extend outside the flower or are hidden inside it. Also look to see whether the blossoms open during the day or night. If you notice this, you can learn to guess what insects, birds, bats, or animals pollinate the flower. It is important also to observe whether the flowers hang and whether they come in clusters, or on upright spikes.
6. If there are no flowers, perhaps there are seed-cases or edible fruits in different shapes, sizes, and colours. Now look under the tree to see if any flowers or fruit have fallen to the ground. Sometimes, you may find seeds of fruit eaten by animals or birds. You can pick up the fallen fruits or flowers and examine them.
7. Notice if the flowers have a scent and whether bees and other insects come for nectar. Look at night to see if moths come to the tree to pollinate the flowers.
8. Note the size, shape, and colour of the leaves. Do you think the tree is deciduous or evergreen? Also, do the leaves grow singly, or are they divided into smaller leaflets? Do the leaves grow alternately up the stem or in pairs?

Note: Watch for all these things and write them down so that you can tell your teachers, friends, or parents. You can keep a notebook for this and you can also press the leaves and flowers of trees between the pages of the book to keep a visual record of what you have seen and learnt. By just observing and noting down such interesting details, you are learning something which will be of life-long benefit to you.

GLOSSARY

The following explanations may help to clear doubts you may have while reading this book. I have put the words in sections according to the parts of the tree and have finished with some general words that may have confused you.

The trunk and growth of a tree

Bole	Another name for the trunk
Branchlets	Small branches of a tree
Buttress	A thickening at the base of the trunk or a set of root supports that keep a tree firmly fixed in the ground
Canopy	The umbrella-like covering of leaves and branches spreading out at the top of the tree
Crown	The top part of a tree where most of the leaves sit
Fissure	A narrow crack or split
Girth	The measurement around the trunk or branches of a tree
Horizontal	When branches or other stems grow out at right angles to the trunk
Indigenous	Of the place, belonging to a country, state or area, native to an area
Longitudinal	Running lengthwise
Vertical	When branches or stems grow up parallel to the trunk
Whorled	Arranged in a circle around a stem

Leaves

Bract	An irregularly-developed leaf lying at the bottom of the flower stalk or of the flower, which is often brightly coloured
Digitate	A leaf with several leaflets radiating from a central point and looking like digits or fingers on a hand
Elliptical	Like an ellipse or regularly oval
Entire	A single leaf with smooth edges

Leaflet	One of several leaf-like growths, which together form a leaf
Lobe	A division of a leaf
Palmate	A leaf that is divided into lobes that look like the palm of our hand
Petiole	Stalk of a leaf
Pinnate	A leaf that is divided into leaflets which lie sideways and are often many in number, thus creating a feathery effect
Serrate-edged	A single leaf with toothed or zig zag edges
Symmetrical	Having exactly the same structure on both sides of the leaf, flower, etc.
Tri-foliate	A leaf made up of three leaflets
Vein	A rib or streak, usually of a different colour on a leaf or flower
Wedge-shaped	Thin at one end and thicker at the other

Flowers

Bisexual	Flowers of both sexes occurring on the same tree
Calyx	The cup into which the petals of a flower sit, often green in colour
Corolla	All the petals of a flower considered as a whole
Fragrance	Perfume or scent of a flower
Petal	A part of the corolla
Petiole	Leaf stalk
Pistil	The female part of the flower consisting of ovary, style, and stigma
Profuse	Many individual units, in this case, flowers or leaves
Sepal	Like petals, these are parts of the calyx that open as the flower develops
Stamen	The male part of the flower consisting of filament and anther
Tassel	A tuft of hanging threads or flowers

Fruit

Fermentation	The breakdown of material or liquid by yeasts or bacteria
Fibre	The threads or filament forming tissue, usually protecting the seed
Pod	A round or long compartment for seeds
Pulp	The fleshy part of a fruit

General terms

Aerial	Above ground, in the air
Antidote	Some chemical taken to counteract a poison
Brace	A support
Degraded	Run down or spoilt

Ecosystems	Communities of plants, animals, or other organisms living together in a particular place or environment
Epiphyte	Any plant that lives on another but takes nourishment from the atmosphere
Erosion	To wear away or destroy the land
Germinate	Showing signs of growth of a new individual
Habitat	The natural home of a plant or animal
Infusion	Leaves, flowers, or other tree products, soaked in water to produce a medicinal liquid
Marine	Any reference to the sea or salt water areas
Narcotic	A substance that affects the mind and changes the way one thinks
Parasite	A plant or animal that lives on and takes food from another plant or animal
Pollinate	To transfer pollen from one flower to another to start fertilization
Species	Class of things having the same characteristics
Tanning	Process of softening and preparing leather

INDEX OF BOTANICAL NAMES

FINAL WORD

You may notice that not all the trees described in this book are illustrated by hand. This is because when updating this book from its earlier edition, I decided to add some trees that I felt were common to India. Since digital photographs have become common, I badgered all my botanical friends to help me with photographs and also took some myself. As a result, some of the trees do not show the flowers and fruit in as much detail as the illustrations do. I hope that this does not detract from the material and that the recognition of the tree will still be easy.